Cellular Interactions in Develo

The Practical Approach Series

SERIES EDITORS

D. RICKWOOD
Department of Biology, University of Essex
Wivenhoe Park, Colchester, Essex CO4 3SQ, UK

B. D. HAMES
Department of Biochemistry and Molecular Biology
University of Leeds, Leeds LS2 9JT, UK

Affinity Chromatography

Anaerobic Microbiology

Animal Cell Culture (2nd Edition)

Animal Virus Pathogenesis

Antibodies I and II

Behavioural Neuroscience

Biochemical Toxicology

Biological Data Analysis

Biological Membranes

Biomechanics—Materials

Biomechanics—Structures and Systems

Biosensors

Carbohydrate Analysis

Cell–Cell Interactions

The Cell Cycle

Cell Growth and Division

Cellular Calcium

Cellular Interactions in Development

Cellular Neurobiology

Centrifugation (2nd Edition)

Clinical Immunology

Computers in Microbiology

Crystallization of Nucleic Acids and Proteins

Cytokines

The Cytoskeleton

Diagnostic Molecular Pathology I and II

Directed Mutagenesis

DNA Cloning I, II, and III

Drosophila

Electron Microscopy in Biology

Electron Microscopy in Molecular Biology

Electrophysiology

Enzyme Assays

Essential Developmental Biology

Essential Molecular Biology I and II

Experimental Neuroanatomy

Fermentation

Flow Cytometry

Gas Chromatography

Gel Electrophoresis of Nucleic Acids (2nd Edition)

Gel Electrophoresis of Proteins
(2nd Edition)
Gene Targeting
Gene Transcription
Genome Analysis
Glycobiology
Growth Factors
Haemopoiesis
Histocompatibility Testing
HPLC of Macromolecules
HPLC of Small Molecules
Human Cytogenetics I and II
(2nd Edition)
Human Genetic Disease
Analysis
Immobilised Cells and
Enzymes
Immunocytochemistry
In Situ Hybridization
Iodinated Density Gradient
Media
Light Microscopy in Biology
Lipid Analysis
Lipid Modification of Proteins
Lipoprotein Analysis
Liposomes
Lymphocytes
Mammalian Cell Biotechnology
Mammalian Development
Medical Bacteriology
Medical Mycology
Microcomputers in Biochemistry
Microcomputers in Biology
Microcomputers in Physiology
Mitochondria
Molecular Genetic Analysis of
Populations

Molecular Imaging in
Neuroscience
Molecular Neurobiology
Molecular Plant Pathology
I and II
Molecular Virology
Monitoring Neuronal Activity
Mutagenicity Testing
Neural Transplantation
Neurochemistry
Neuronal Cell Lines
NMR of Biological
Macromolecules
Nucleic Acid and Protein
Sequence Analysis
Nucleic Acid Hybridisation
Nucleic Acids Sequencing
Oligonucleotides and
Analogues
Oligonucleotide Synthesis
PCR
Peptide Hormone Action
Peptide Hormone Secretion
Photosynthesis: Energy
Transduction
Plant Cell Culture
Plant Molecular Biology
Plasmids
Pollination Ecology
Postimplantation Mammalian
Embryos
Preparative Centrifugation
Prostaglandins and Related
Substances
Protein Architecture
Protein Engineering
Protein Function

Cellular Interactions in Development

A Practical Approach

Edited by

DAVID A. HARTLEY

Department of Biochemistry,
Imperial College,
London SW7 2AZ

IRL PRESS
at
OXFORD UNIVERSITY PRESS
Oxford New York Tokyo

Oxford University Press, Walton Street, Oxford OX2 6DP
Oxford   New York   Toronto
Delhi   Bombay   Calcutta   Madras   Karachi
Kuala Lumpur   Singapore   Hong Kong   Tokyo
Nairobi   Dar es Salaam   Cape Town
Melbourne   Auckland   Madrid
and associated companies in
Berlin   Ibadan

Oxford is a trade mark of Oxford University Press

A Practical Approach 🔾 is a registered trade mark
of the Chancellor, Masters, and Scholars of the University of Oxford
trading as Oxford University Press

Published in the United States
by Oxford University Press Inc., New York

A catalogue record for this book is available from the British Library

Library of Congress Cataloging in Publication Data
Cellular interactions in development : a practical approach / edited
by David A. Hartley.
(The Practical approach series)
Includes bibliographical references and index.
1. Cell interaction—Research—Methodology.   2. Developmental
cytology—Research—Methodology.   I. Hartley, David A.   II. Series.
QH604.2.C454   1993   574.87'61—dc20   93–10301
ISBN 0–19–963391–6 (h/b)
0–19–963390–8 (p/b)

Typeset by Footnote Graphics, Warminster, Wilts
Printed in Great Britain by Information Press Ltd, Eynsham, Oxon

Preface

It has become apparent over the years, that in order to correctly form the diverse cell types of an adult organism from a fertilized egg, cells must interact with each other. Even in organisms which were thought to proceed through development solely by a lineage-dependent mechanism (such as *Caenorhabditis elegans*), cellular interaction has been shown to be necessary to segregate some cell types from each other. This is immensely reassuring to those of us working on *Drosophila*, since it demonstrates the fundamental conservation of cellular interaction mechanisms and the molecules which encode them. It also lets us make grandiose claims of significance in our grant proposals. At a first glance, it might seem impossible to unite a set of experimental protocols under this umbrella, and so this volume represents a personal bias of the most powerful techniques available for understanding the importance of cellular interactions and dissecting them at the molecular level. Inevitably, this leads to a systems bias, particularly in the more biological experimental approaches. However, I hope that those researchers will be encouraged to read chapters which are not about their favourite organism, to understand the general importance of the *in vivo* approach, and to apply techniques with such potential as, for example, inducible somatic mosaicism, or targeted cell ablation.

In this volume of the *Practical Approach* series, I have drawn together a number of diverse methods of general significance in unravelling the mysteries of cellular communication. They fall into two broad categories—firstly those methods which are used to demonstrate the existence of cellular interactions during development, and secondly those methods which are used to isolate and assay the molecules involved in mediating cellular interaction. The first two chapters deal with perhaps the most powerful technique for analysing the effect of one cell population on another during development—by making genetic mosaics. Although these chapters are specific to *Drosophila* and plants, the potential of this technique applies to any experimental system. Chapter 3 covers cell transplantation in detail in *Drosophila*, another method for creating a mosaic individual so that the developmental potential, and influence, of cells can be measured. The genetic approach to development works on the basis that the organismal consequences of ablation of a single gene tells us about the natural function of that gene. Similarly, Chapter 4 describes a technique for targeting cellular ablation, from which one can infer the natural function of the targeted cell population in influencing the development of its neighbours. The remaining chapters are generally applicable methods for examining protein–protein interaction and assaying factors which influence development. Chapter 5 describes ways of cloning molecules which might include the cell surface component of inductive or interactive events, starting from an antibody, or a ligand to an unknown receptor (or vice versa). Chapter 6 includes a general discussion and protocols for analysing protein–protein interaction, which is the key to identifying all the members of a cell interaction pathway. An elegant alternative approach is presented in Chapter 7 which describes a way of analysing protein–protein interaction by a molecular genetic approach. Finally, Chapter 8 describes protocols for analysing mesoderm-inducing factors in *Xenopus*, which has provided a rich source for identifying potential *in vivo* function of members of the FGF, TGF-β, and wnt family of secreted growth factors.

I am indebted to all the authors for their sterling efforts, and to Dr Helen Flinn for goading me into finishing, even when my boiler blew up.

London D.H.
February 1993

Contents

3. Clonal analysis in plants

59

Mark Dudley and R. Scott Poethig

4. Toxin ablation in *Drosophila*

Hermann Steller

5. Cloning cell surface molecules by transient expression in mammalian cells

David L. Simmons

8. Purifying and assaying mesoderm-inducing factors from vertebrate embryos 181

Jim C. Smith

Appendix: Suppliers of specialist items 205

Index 207

Contributors

PAUL L. BARTEL
Department of Microbiology, State University of New York at Stony Brook, Stony Brook, New York 11794, USA.

CHENG-TING CHIEN
Department of Biochemistry and Cell Biology, State University of New York at Stony Brook, Stony Brook, New York 11794, USA.

MARK DUDLEY
Plant Science Institute, Department of Biology, University of Pennsylvania, Philadelphia, PA 19104-6018, USA.

STANLEY FIELDS
Department of Microbiology, State University of New York at Stony Brook, Stony Brook, New York 11794, USA.

KENT G. GOLIC
Department of Biology, 201 Biology Building, University of Utah, Salt Lake City, UT 84112, USA.

DAVID P. LANE
Cancer Research Campaign Laboratories, Cell Transformation Research Group, Department of Biochemistry, University of Dundee, Dundee DD1 4HN, UK.

CAROL A. MIDGLEY
Cancer Research Campaign Laboratories, Cell Transformation Research Group, Department of Biochemistry, University of Dundee, Dundee DD1 4HN, UK.

R. SCOTT POETHIG
Plant Science Institute, Department of Biology, University of Pennsylvania, Philadelphia, PA 19104-6018, USA.

ANDREAS PROKOP
Universität Mainz, Institut für Genetik-Zellbiologie, Saarstraße 21, 6500 Mainz, Germany.

DAVID L. SIMMONS
Cell Adhesion Laboratory, Imperial Cancer Research Fund Laboratories, Institute of Molecular Medicine, John Radcliffe Hospital, Headington, Oxford OX3 9DU, UK.

JIM C. SMITH
Laboratory of Developmental Biology, National Institute for Medical Research, The Ridgeway, Mill Hill, London NW7 1AA, UK.

Contributors

HERMANN STELLER
Howard Hughes Medical Institute, Department of Brain and Cognitive Sciences, Massachusetts Institute of Technology, Cambridge, MA 02139, USA.

ROLF STERNGLANZ
Department of Biochemistry and Cell Biology, State University of New York at Stony Brook, Stony Brook, New York 11794, USA.

GERHARD M. TECHNAU
Universität Mainz, Institut für Genetik-Zellbiologie, Saarstraße 21, 6500 Mainz, Germany.

Abbreviations

AEM	acrylamide embedding medium
AUT	autonomy undergoing testing
BSA	bovine serum albumin
CHAPS	3-[(3-cholamidopropyl)-dimethylammonio]-1-propane-sulfonate
DAB	diaminobenzidine
DABCO	1,4-diazobicylo-[2.2.2]-octane
DAPI	4′,6-diamino-2-phenylindol-dihydrochloride
DEAE	diethylaminoethyl
DEPC	diethylpyrocarbonate
DME	Dulbecco's modified Eagle's medium
DOC	deoxycholic acid
DT	diphtheria toxin
DTT	dithiothreitol
EL	egg length
EM	electron microscope
FGF	fibroblast growth factor
FITC	fluorescein isothiocyanate
FRT	FLP recombination target
GST	glutathione-S-transferase
GuSCN	guanidinium isothiocyanate
HCG	human chorionic gonadotrophin
HPLC	high-performance liquid chromatography
HRP	horseradish peroxidase
HSP	heat-shock protein
L1	first cell layer of the shoot apical meristem
L2	second cell layer of the shoot apical meristem
L3	third cell layer of the shoot apical meristem
LM	light microscope
LPA	linear polyacrylamide
M	marker
Mab	monoclonal antibody
NAM	normal amphibian medium
NBT	p-nitroblue tetrazolium chloride
pA+	poly-A enriched RNA fraction
PB	phosphate buffer
PBS	phosphate-buffered saline
PEG	polyethylene glycol
PMSF	phenylmethylsulfonyl fluoride
PMSG	pregnant mare serum gonadotrophin

R	roentgen
RACs	ricin A-chain cold sensitive mutant
RB	retinoblastoma
RITC	rhodamine isothiocyanate
SAM	shoot apical meristem
SDS	sodium dodecyl sulfate
TCR	T-cell receptor
TEMED	N,N,N',N'-tetramethylethylenediamine
TFA	trifluoroacetic acid
TMB	3,3',5-5'-tetramethylbenzidine
USCE	unequal sister chromatic exchange
VD	ventrodorsal

1

Generating mosaics by site-specific recombination

KENT G. GOLIC

1. Introduction

Does the developmental fate assumed by a cell depend on signals that it received from its neighbouring cells? This is the fundamental question addressed by the study of cellular interaction during development. If cell–cell communication can be implicated in the specification of cell fates then a series of further questions can be asked. Which cells are talking and which are listening? What molecules are involved and how do they work? When are these molecules needed? In organisms in which there are mutations available that alter cell fate these questions can be fruitfully investigated by the analysis of mosaics. There are molecular techniques available that can tell when and where a gene is expressed; genetic mosaics can be used to determine which part of this expression pattern is functionally significant (see also Dudley and Poethig, this volume).

A mosaic individual is one in which cells or groups of cells can be distinguished as having a separate origin. Most commonly this term refers to genetic mosaics in which cells of different genotypes develop together in the same individual. A specific method for efficiently generating such mosaics is the topic of this chapter. In brief, at almost any stage of development a subset of cells in an organism can be altered genetically, so that at a later stage of development the mitotic progeny of such cells can be distinguished and their fate determined.

The principle of using mosaics to ask questions about cell–cell interaction is straightforward. Suppose that a gene is suspected of involvement in cell-to-cell signalling that specifies a certain cell fate. By producing individuals that are mosaics of mutant and wild-type cells one can ask which cells must be genetically wild-type in order for the cell of interest to assume its proper fate. If the gene is involved in sending a signal then this cell may be mutant and still assume its appropriate fate. In this case the gene is said to function in a nonautonomous fashion because the fate of a cell does not depend on its own genotype at this locus. If the gene is involved in receiving a signal then when the cell of interest lacks this receiver it would fail to assume its proper fate. Here, the gene is said to function autonomously. By varying the developmental stage of mosaic induction one can ask when the gene must be active in order to perform its function. If the gene is indeed involved in a signalling process then it should function at the time the choice of fates is made or before. The advent of the techniques discussed here has simplified the production of mosaic individuals, making it relatively easy to generate them for these analyses.

As a first step towards determining whether the specification of a cell's fate requires cell–cell interactions it can be worthwhile to show that lineage alone cannot account for the determination of a cell's fate. Through the use of animals that were mosaics for a gratuitous cell marker, Ready, Hanson, and Benzer demonstrated that the eight photoreceptor cells of a single ommatidium in the *Drosophila* compound eye are not clonally related (1). It has since become clear that cell fates in the compound eye are specified through a process that involves extensive cellular communication (2). Even if clonal analysis does indicate that cells follow an invariant lineage, such a finding does not exclude the contribution of cellular interactions. Witness *Caenorhabditis elegans*—it is now evident that at least some of the invariant cell lineages that produce this animal are maintained through cell–cell interactions (3).

The technique of mosaic analysis provides several advantages for studying cellular communications. Principal among these are that existing mutations can be analysed to sort out the contributions of different genes, the cells develop *in vivo* so their physiology is optimal, and cell fate can be assayed in the normal developmental context. Deviations from the normal fate reflect alterations in the genotype of the cell of interest or the cells with which it interacts. It is of prime importance then to know which cells in a mosaic animal are mutant and which are wild-type. This is usually accomplished by a gratuitous marking of the cell that occurs simultaneously with the loss (or possibly gain) of the gene function in question.

2. Generating mosaics by site-specific recombination

A number of techniques have been developed for generating mosaics, each with its own set of advantages and drawbacks (4–9). The technique of site-specific genetic recombination is probably the most broadly applicable. It works in a variety of organisms, its suitability to *in vitro* manipulation allows for customized design of the desired mosaics, and it is extremely efficient. In *Drosophila melanogaster* one can readily arrange for virtually every animal in an experiment to become mosaic. Because of this great efficiency many mosaics can be analysed and conclusions are placed on a solid basis of multiple observations; in other words, one can feel comfortable with the conclusions drawn.

Site-specific recombination systems typically consist of a limited number of proteins that recognize specific DNA sequences and effect genetic recombination between them (10). The simplest systems, and thus most amenable to manipulation, are those that comprise just a single protein recombinase and the sequence that it recognizes. In theory, when a recombinase gene is placed under appropriate transcriptional and translational control for the organism that is being studied, and the recognition sites integrated in the genome of that organism, then synthesis of the active recombinase will mediate recombination between the embedded sequences. This then leads to an alteration in the genotype of a cell and its daughters. Two simple systems that have been transferred to a variety of organisms, where they behave much as expected, are the *FLP–FRT* system from the *Saccharomyces cerevisiae* 2μ plasmid (11, 12) and the *Cre-lox* system from bacteriophage P1 (13).

2.1 *FLP–FRT* site-specific recombination system

The 2μ plasmid encodes a 48 kDa recombinase called FLP, and carries two copies of the sequence that FLP recognizes, called *FRT*s (FLP Recombination Target). The arrangement of *FRT*s in relation to each other and to the origin of replication enables the plasmid to use FLP-mediated recombination during the S-phase to shift into a type of rolling-circle replication and increase in copy number.

2.1.1 The recombinase

FLP requires no other yeast genes to mediate recombination. The protein has been synthesized in *Escherichia coli* and purified to homogeneity, and is capable of carrying out recombination *in vitro* without a high energy cofactor or divalent cations.

The active form of the FLP protein appears to be short-lived—an advantage for mosaic studies where it is frequently important to control the timing of marking cells. The purified protein is very unstable *in vitro* under conditions that optimize its activity; FLP protein loses its activity completely after half an hour in the absence of substrates. When the *FLP* gene is expressed in *Drosophila* its mRNA is probably subject to rapid degradation. When the activities of two heat-shock promoter-driven *FLP* genes were compared, one that made an mRNA with the *hsp70* 3′ end and one with the *FLP* 3′ end, that with the *hsp70* 3′ end showed significantly higher levels of activity both with and without heat shock (R. Petersen and K. Golic, unpublished data). The 3′ noncoding end of the *Drosophila hsp70* mRNA confers extreme instability to that message in cells that have not been stressed by heat (14). The observation that a message with the *hsp70* 3′ end makes more FLP than one with the *FLP* 3′ end suggests that the *FLP* 3′ end may target its message for especially rapid degradation. This instability in *Drosophila* may be merely a happy circumstance. However, in yeast the *FLP* gene is tightly regulated. Several genes act in concert to control *FLP* transcription. A reasonable extension of these observations is to suppose that selective pressures that have led to tight controls over transcription, and possibly translation, have also acted to make the protein short-lived. After all, it makes little sense to have tight controls over the synthesis of a protein if it is not turned over rapidly. Although it might be possible to further ensure rapid degradation of FLP by fusing the gene to a sequence coding for a series of amino acids that would confer instability, the few terminal amino acids at both ends of the protein are required for recombinase activity (15): a peptide fused to either end might very well destroy its ability to effect recombination.

2.1.2 The *FRT*

The 2µ plasmid contains two precise copies of a 599 base pair (bp) sequence that lie almost directly opposite each other on the plasmid in inverted orientation. The sequences recognized by FLP and required for recombination (*Figure 1*) occupy the centre of each repeat about an XbaI restriction site. The XbaI site lies in what is called an 8 bp spacer region flanked by 13 bp inverted repeats (with a single base mismatch that is apparently without consequence). FLP binds to the outside base of the spacer and the 11 most proximal bases of each inverted repeat. The 13 bp sequence is repeated a third time (with a single intervening base) but this repeat is not required for recombination. The

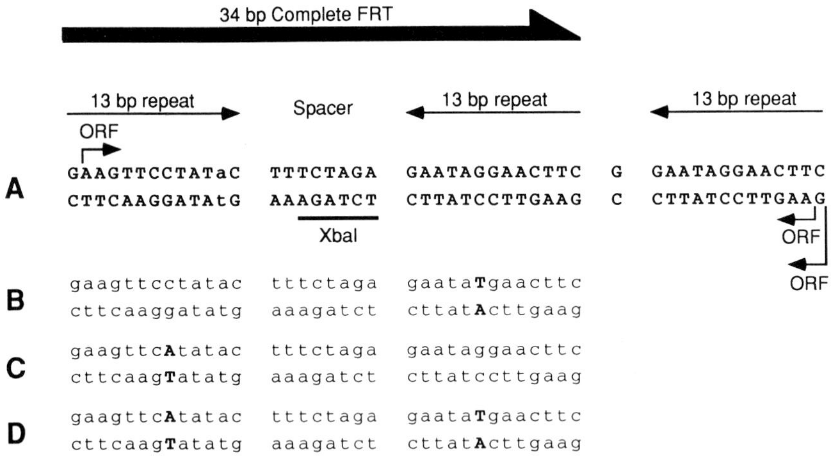

Figure 1. The nucleotide sequence of the *FRT*. Features indicated are discussed in the text. (A) The wild-type sequence; the first 13 bp repeat differs from the other two by a single base change, indicated in lower case letters. (B) The sequence of a mutant *FRT* that reduces recombination efficiency by a factor of five; the mutation is indicated in upper case letters. (C) The sequence of the same mutation in the other 13 bp repeat; it also reduces recombination efficiency five-fold. (D) The sequence of the double-mutant that reduces recombination efficiency 100-fold. ORF = Open Reading Frame.

FRT has been extensively characterized *in vitro*. Only 28 bp, the central spacer and 10 bp of each repeat, are required for full recombination activity using this assay, although sequence gazing would suggest that a complete *FRT* consists of at least 34 bp: the central spacer and two complete inverted repeats.

FLP makes staggered single-stranded cuts at the 5′ ends of the spacer and becomes covalently attached to the DNA via a transient phosphotyrosyl linkage. Recombination proceeds through a Holliday intermediate that FLP can resolve to generate recombinant molecules. The spacer region possesses an asymmetry that gives the *FRT*s, and thus the recombination reaction, directionality. (Other asymmetrical elements in this region, the single base mismatch and the third 13 bp repeat, are not involved in generating directionality.) The XbaI site is offset from the centre of this 8 bp. We use the convention of indicating *FRT* direction with an arrow, the head of which is closer to the XbaI end of the spacer. If the spacer region is made symmetrical by shifting the XbaI site by one base pair, recombination proceeds efficiently *in vitro* but the directionality of the reaction is abolished. The outcome of a recombination event between two asymmetrical *FRT*s can be predicted on paper by aligning the *FRT*s above one another so that the spacer regions are in the same orientation and drawing a simple reciprocal exchange event between them (*Figure 2*). Intramolecular recombination between *FRT*s in the same orientation will lead to the excision of a circular molecule that carries one *FRT* and all the DNA that lay between the two *FRT*s. One *FRT* remains behind. If the *FRT*s are inverted then FLP-mediated recombination between them will invert the intervening DNA. FLP also catalyses intermolecular recombination with results that vary according to the orientation and position of the *FRT*s, and the relationship of the recombining DNAs.

Although the spacer sequence determines the directionality of recombination, the exact composition of the spacer matters little. Single base pair changes in the central six bases have little effect on recombination efficiency as long as the spacers of recombining molecules match. This tolerance of sequence alteration, but requirement for homology, has provided a method for directing recombination to occur between selected pairs of *FRT*s *in vitro* when multiple *FRT*s are present. Some multiple base substitutions in the spacer region are not tolerated: in particular, changes that increase the GC content of the spacer or disrupt the string of purines extending from the spacer into the 13 bp repeat.

Single base changes in the 13 bp repeats have also been examined and have variable effects on recombination efficiency *in vitro*. One pair of mutations is of particular interest. Mutating a specific G to T (see *Figure 1*) in one of the two 13 bp repeats of a 34 bp *FRT*, reduces five-fold the recombination efficiency observed between this *FRT* and another similarly mutated *FRT*. If, however, both 13 bp repeats of a single *FRT* are thus mutated then recombination efficiency between this double-mutant *FRT* and another wild-type or

<u>Direct Repeats</u>

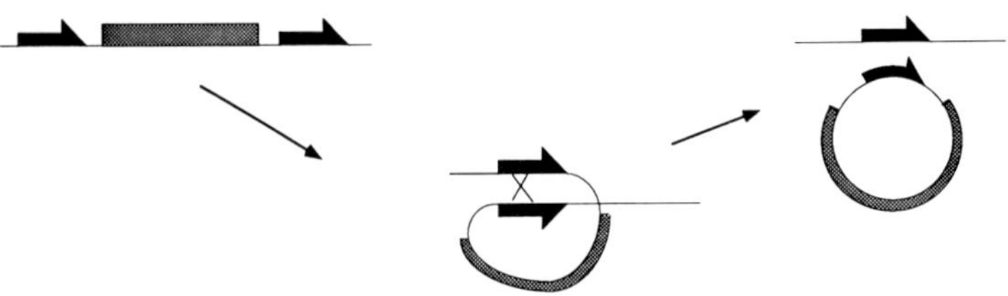

<u>Inverted Repeats</u>

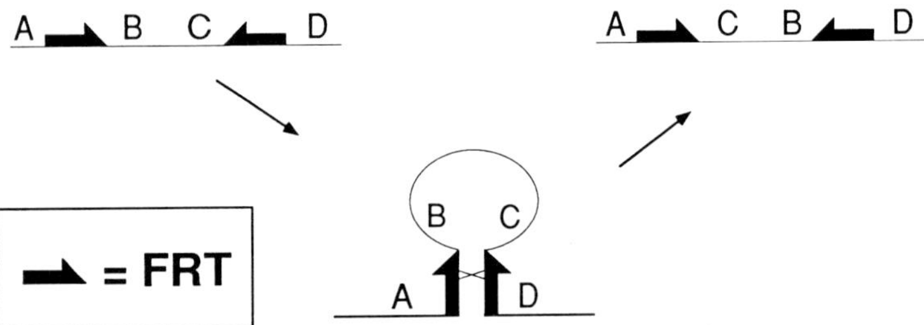

Figure 2. The consequences of intramolecular exchange between *FRT*s.

mutant *FRT* is reduced more than 50-fold. When two *FRT*s recombine the product *FRT*s are actually composites with the repeats on either side of the spacer derived from different parental *FRT*s. By arranging two single-mutant *FRT*s so that recombination between them produces one double-mutant *FRT* (and one wild-type *FRT*), one might effectively inactivate an *FRT* and 'lock' the recombination product in a configuration that would ordinarily be reversed by further FLP-mediated recombination. The effectiveness of this method has not been demonstrated *in vivo*. One test using a bacterial model indicated that the single-mutant *FRT*s reduced recombination frequencies far more than the *in vitro* results had suggested (16). When tested in mosquito embryos these mutant *FRT*s were somewhat effective, but at a lower level than predicted (17). The level of FLP may be critical in these experiments and the use of such mutated *FRT*s is certainly worth further exploration.

2.2 *Cre-lox* site-specific recombination

A second site-specific recombination system that has been utilized as a genetic tool is the *Cre-lox* system from bacteriophage P1. Cre is a 38 kDa protein that mediates recombination between 34 bp *lox* sites. Its normal function, to effect resolution of the P1 lysogen plasmid multimers, stabilizes the prophage by increasing the number of segregating units. This system is very similar to the *FLP–FRT* system: the recombination targets are directional, recombination can occur between inverted and direct repeats, intramolecular and intermolecular reactions occur, Cre is the only protein necessary for recombination, and it has been shown to work in several other species. The *lox* recombination sites, like *FRT*s, consist of 13 bp inverted repeats flanking an 8 bp asymmetrical spacer. Recombination using Cre on integrated *lox* sites has not been as efficient in most cases as FLP-mediated recombination on integrated *FRT*s, but that may be a function of the promoters used to express the recombinases (see Section 3.1.3*ii*).

2.3 Other systems

Other site-specific recombination systems have also been used across species boundaries. Many yeasts contain 2μ-like plasmids that encode recombinases. One of these, from *Zygosaccharomyces rouxii*, can function in tobacco cells (18). The Gin recombinase of bacteriophage Mu has been transferred to plant protoplasts where it catalyses site-specific recombination (19). The wild-type Gin recombinase normally requires an *E. coli* accessory protein (FIS) and catalyses only intramolecular inversions within phage Mu that control its host range. In these plant cells a FIS-independent mutant Gin protein was used, bypassing the need for any additional proteins. This mutant Gin is also capable of catalysing recombination between directly oriented target sites (*gix*) and intermolecular recombination.

2.4 Fundamental properties

The following discussions focus mainly on the *FLP–FRT* system in *Drosophila melanogaster* because the use of site-specific recombination on chromosomally located target sites has been explored most thoroughly in this organism. In most other organisms (with the exception of tobacco plants) site-specific recombination has so far been used only in cultured cells or on extrachromosomal target sites. However, the ideas and techniques discussed should be broadly applicable when whole organism methods are developed in other species and with other recombinases.

The initial demonstration of site-specific recombination in *Drosophila* utilized two simple constructs that were placed into the genome by P element mediated transformation (20). First, a heat-inducible FLP gene (*hsFLP*) was constructed by joining the *Drosophila hsp70* promoter and 5′ transcribed but untranslated leader sequence to the *FLP* coding sequence and 3′ untranslated region. An assay for FLP activity was provided by cloning a cell-autonomous marker gene between directly repeated *FRT*s (as in *Figure 2*). The marker gene chosen was an allele of the *white* gene (w^{hs}), which is needed for deposition of the reddish pigments that characterize the wild-type eye. This construct was called P[>w^{hs}>] where the > symbol serves as a shorthand designation for an *FRT* and the brackets indicate the material carried by the P element. Chromosomal insertions of these constructs were crossed into a genetic background where the normal copy of the *white* gene carried a null mutation (w^{1118}). Larvae that carried both insertions were briefly heat-shocked to induce FLP synthesis, which was expected to cause the excision and loss of the w^{hs} gene. The adults that eclosed were inspected for eye colour mosaicism. The results showed that FLP can work with great efficiency on *FRT*s integrated into the chromosomes of *Drosophila*. A single heat shock could induce excision of the *FRT*-flanked w^{hs} gene in more than 90% of the cells in the soma as assessed by eye-colour mosaicism, and in the male and female germlines as demonstrated by mating the mosaics and scoring their progeny for the presence of the w^{hs} gene. In these experiments every heat-shocked animal exhibited somatic mosaicism, further illustrating the extreme efficiency of the 2μ system. This technique, and extensions of it, are used to generate individuals that are mosaic for genes that have been cloned.

Site-specific recombination between chromosomally located target sites has also been used to generate mosaics of genes identified by mutation only (21). FLP readily catalyses recombination between allelic *FRT*s, that is, *FRT*s inserted at the same site on homologous chromosomes. If the chromosomes are differentially marked with alleles of a gene distal to the site of *FRT* insertion then the mitotic recombination that is catalysed by FLP can produce daughter cells that are homozygous for each of the alleles (as shown in *Figure 3*). If recombination is induced early in development then these daughter cells can proliferate to produce genetically marked clones. Like the method of excision, mitotic recombination mediated by FLP is extremely efficient, making mosaics of virtually all individuals after FLP induction.

2.4.1 A caution

The efficiency shown by site-specific recombination is in some ways a mixed blessing. It makes it easy to generate individuals that are mosaic in virtually any tissue that one wishes to examine. The cost is that such individuals may have several independently generated clones. These independent clones can particularly frustrate lineage analyses and, to a lesser extent, confuse the interpretation of mosaics when attempting to elucidate cell–cell interactions. Fortunately these problems can be dealt with rather easily. First, one can adjust the frequency of mosaicism so that multiple events are infrequent (Sections 3.1.1*i,ii* and 3.3.1). Second, many mosaics can be examined so that when there is a change in the fate of a cell, the contribution of clones generated in the tissue of interest can be sorted out from the contribution of clones generated elsewhere.

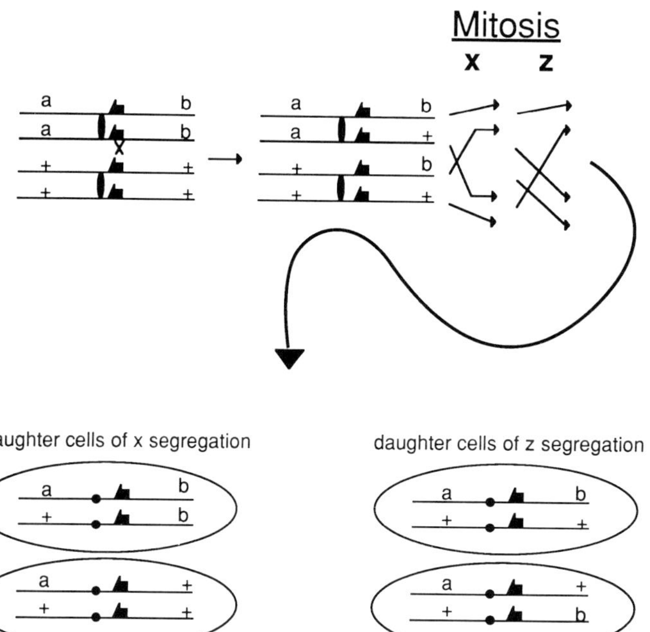

Figure 3. Mitotic recombination at *FRT*s. Shown here are the possible outcomes following a single exchange between nonsister chromatids in G2 of the cell cycle. The solid ovals and circles indicate the centromeres.

3. Practical aspects

3.1 Expression of recombinase

3.1.1 Heat shock

i. Technical details

The *hsp70* promoter-driven *FLP* gene is a very good general purpose source of FLP for use in *D. melanogaster*. By varying the severity of heat shock it can be easily induced to give varying levels of recombination at most stages of development and in most tissues. The manner of heat-shock application can greatly influence the frequency of recombination that is achieved, presumably because more FLP is expected to be synthesized after more severe heat shocks. Higher temperatures or longer times increase the resulting frequency of recombination. When the method given in *Protocol 1* is followed, the maximum heat shock that can be applied is 38.5 °C for one hour. Little or no larval death results from this treatment (as shown by the presence of melanotic dead larvae in the food), but raising the temperature to 39 °C kills virtually all larvae and pupae in the vial. Digitally-controlled heating circulators with a control accuracy of ±0.1 °C or better are most convenient to use because of the ease and rapidity with which the temperature can be changed. Some examples are (and there are many others) Braun Thermomix B, Lauda MS, and Cole-Parmer 1253-00.

Higher temperatures and shorter times could presumably be used but we usually avoid heat-shocking for less than one hour. It takes approximately 8–10 minutes for the centre of the food in a vial to rise from 25°C to within 0.5°C of a target temperature of 37°C when the vials are immersed in a circulating water bath as described. If a short heat shock were to be used then larvae occupying an interior position might experience a substantially less severe treatment than larvae closer to the walls of the vial. A longer heat treatment reduces the relative magnitude of this larva to larva variation.

Heat shocks given in circulating water baths are much more effective at inducing recombination than are heat shocks of the same duration given in air incubators. Heat transfer is much swifter in water and the interior of the vial reaches the target temperature much more quickly. In an air incubator (without fan) the centre of the food in a vial takes an hour or more to rise to within 0.5°C of the target temperature of 37°C. Pretreating cells at a moderate temperature protects them from the effects of subsequent higher temperatures and allows them to respond to a given temperature with a lower level of heat-shock protein synthesis than if they had not been pre-treated (22). It is likely that the slow rate of warming in an air incubator presents a similar situation. The time the larvae spend at the maximum temperature is reduced, and because they are heated slowly they have time to mount an initial response before maximum temperature is attained. The heat-shock proteins synthesized at the lower temperatures probably prevent the larvae from synthesizing as much *hsp70*-promoter driven FLP as they would have had they been heated quickly. If heat shocks must be performed in an air incubator I recommend at least doubling the time. If an incubator of sufficient size is used, quick heating may be facilitated by prewarming a pan of water in the incubator and placing the rack of vials in this pan of water. When using an air incubator it may be more convenient to vary the severity of heat shock by changing its duration rather than its temperature.

Small differences in the temperature of heat shock also make a noticeable difference in the frequency of recombination. A heat shock of one hour at 38.5°C produces roughly twice the frequency of somatic excision as one hour at 37°C. I recommend that a series of standard heat shocks (*Protocol 1*) be performed at various temperatures and the resulting variation in mosaicism observed. This will help give a feeling for the frequency of mosaicism that may be obtained with the equipment at hand and the sizes and shapes of clones produced by recombination at various stages. In a circulating water bath heat shocks of one hour at 36°C, one hour at 37°C, or one hour at 38.5°C should provide a good range of excision frequencies. Finally, in experiments where adults are to be heat-shocked they should first be allowed to fully recover from anaesthesia.

Protocol 1. Heat shocking in a circulating water bath

1. Mate 2–4 pairs of *w^{1118}; hsFLP2B/CyO* males × *w^{1118};* P[>*w^{hs}*>] virgin females in 25 × 100 mm cotton-plugged glass shell vials on standard food (20–25 mm deep).

2. Allow them to lay eggs at 25 °C until pupae appear on the sides of the vials (5–6 days). Remove parents—they may be transferred to a new vial to repeat the procedure.

3. Adjust a circulating water bath to 37 °C (or other desired temperature) and place an appropriately sized test tube rack 75 mm deep in the bath. When the water bath reaches temperature place the vial that contains the progeny into the rack and then push down the cotton-plug at least far enough so that the bottom of the plug is slightly below the plane of the water's surface. In this way it can act as an insulator and ensure that larvae do not crawl up to a cooler region.

4. After one hour remove the vial and allow to cool on the bench or in the 25 °C incubator. If several vials are heated they should remain separated so that air can circulate around them while cooling. It is best to decide on a standard protocol for cooling and then stick with it. Putting warm vials into a box where they may contact other vials should be avoided.

5. When flies begin to eclose in 4–5 days they should be collected and examined daily for mosaicism. Each succeeding day will produce flies that were heat-shocked approximately 24 hours earlier in their life cycle.

One additional method of heat-shocking is worth mentioning. Monsma *et al.* (23) have used a technique whereby they could limit the heat-shocked area to just a small portion of a *Drosophila* embryo. A heated probe is brought into contact with the surface of an embryo at cellular blastoderm—the resulting heat-shock response is limited to just a few cells. This technique might be useful to target mosaicism to the progeny of a few specifically chosen cells.

ii. Position effects on hsFLP activity

The expression of transformed copies of *hsFLP*, like that of most genes, is influenced by their chromosomal environment. This necessitates the characterization of different *hsFLP* insertions with respect to their constitutive level of activity and degree of inducibility in the tissue of interest. One insertion of *hsFLP* that we have used (*hsFLP*2B) is strongly inducible in virtually all parts of the body that we have examined and shows much lower levels of constitutive activity in those tissues. For example, when the eye-colour mosaicism that results from excision of the *FRT*-flanked *w^{hs}* gene was examined, virtually all animals became mosaic after a single one-hour heat shock of 37 °C was given during the first five days of development. Without heat shock about 10–15%

of such animals were mosaic (20). The difference in the frequency of recombination events is actually much greater than it appears; about half the cells in the eyes of the flies given this heat shock experienced excision events, but without heat shock the mosaics that arose most often consisted of a single clone. Whether these mosaics resulted because this insertion expresses at a low level without heat shock or because some larvae actually experienced a low level of heat-shock promoter-inducing stress (such as anoxia) is unknown. It is clear, though, that some *hsFLP* insertions do have demonstrably high constitutive levels of activity. The insertion *hsFLP*1 has lower constitutive and inducible levels of activity than *hsFLP*2B in the eye and the germline, but in the notum has a very high constitutive level (21). Much of the time this high constitutive level will be undesirable, but it can serve to provide high levels of mosaicism without the need to heat shock, and may thus be useful in some experiments.

A second heat-inducible *FLP* construct (called *70FLP*) has a high constitutive level of activity, as determined by excision of an *FRT*-flanked white gene, when inserted at any of several locations throughout the genome. The activity of this gene in the eye without heat shock is roughly equivalent to *hsFLP*2B after a 37°C one hour heat shock. In contrast, the constitutive activity of this gene in the male and female germlines produces only a few per cent recombination without heat shock. The *70FLP* gene is also strongly induced by heat shock.

The varying levels of activity in different tissues with different insertions of the same gene or insertions of different heat-inducible constructs points out the importance of characterizing the activity of the chosen FLP gene in the context of the experiment that one has planned.

iii. Cells that are refractory to heat shock

It is worth considering whether all cells at all stages of development respond to heat shocks; most do. Cuticle markers have been used to show that essentially all parts of the body are susceptible to FLP-induced mitotic recombination after heat-shock is used to induce FLP synthesis (21), and R. Holmgren (personal communication) has used a FLP-activated *lacZ* construct to mark internal tissues and found, to a first approximation, that all are susceptible to such marking when *hsFLP* is used.

There are some notable exceptions. First, preblastoderm nuclei do not respond to heat shock. This probably accounts for the high lethality induced by heat treatments of this early stage. Only embryos that have reached the blastoderm stage respond to heat shocks by inducing the expression of heat-shock proteins (24).

A second exception is the post-stem cell stages in the male germline. Bonner *et al.* (25) showed that an *hsp70–ADH* fusion gene could be induced in the stem cells of the testis but not in primary spermatocytes. Results obtained by mating adult males that were heat-shocked at various stages of larval development and assaying the chromosomes carried by their gametes for FLP-mediated recombination are consistent with this (unpublished data). In order, then, to maximize the fraction of gametes arising from cells in which recombination occurred, males should be heat-shocked early in development —within the first day or two after egg deposition. Males heat-shocked later than this show lower levels of germline recombination, presumably because many of the gametes sampled derive from spermatogenic stages that did not respond to the heat shock. Adult males may be heat-shocked, but one should not expect to recover high frequencies of recombinant gametes for 9–10 days afterwards (when flies are raised at 25 °C). In the female germline, high levels of recombination can be achieved until at least two days before eclosion. Higher levels of recombination are achieved when heat shocks are applied during the pupal stage than when they are given to larval stages, probably because there are more germ cells present in these later stages. Females heat-shocked as adults should produce mature recombinant oocytes within a short time.

A third exception exists in the eye discs of third-instar larvae. When third-instar larvae carrying a heat-inducible *lacZ* gene are heat-shocked and their eye discs stained with X-gal, it can be seen that the cells of the eye disc do not respond uniformly to heat shock. Cells that lie in a vertical stripe just anterior to the morphogenetic furrow respond weakly to heat shocks, while cells just behind the furrow respond quite strongly (unpublished data). These relative responses are reflected in the patterns of mosaicism that result when *hsFLP* is induced at this stage (see Section 3.2.2i).

iv. Heat-induced phenocopies

Heat shock can induce developmental aberrations that mimic the phenotypes caused by some mutations (4)—including some mutations that are commonly used as cuticle markers for mitotic recombination. Fortunately these phenocopies can only be induced during specific developmental stages which tend to be much later (usually during the pupal stage) than one would typically heat shock to induce recombination.

v. Other forms of stress and developmental induction of heat shock promoters

The *hsp70* promoter can also be induced by agents other than heat, including anoxia, heavy metals, compounds that inhibit respiration, and many more (26). Developmental induction of *hsp70* in *Drosophila* has not been reported, but other heat shock genes are expressed in the course of normal development.

3.1.2 Other methods of expression

The expression of FLP need not of course be limited to transcription driven by a heat-shock promoter. Another useful method might be to join this gene to a developmentally regulated promoter. A versatile variation would be to have *FLP* expression driven by GAL4 enhancer traps (see Chapter 4).

3.1.3 Expression in other organisms

i. *Poly-A signals in the* FLP *gene*

When considering other constructs, especially for expression in other organisms, it should be noted that the *FLP* gene carries three consensus polyadenylation signalling sequences within its protein-coding region. These poly-A signals have not prevented the robust expression of *FLP* after heat shock in *Drosophila*, but they may have interfered with the expression of an *FLP* construct in mammalian cells. Heat shocks disrupt the normal transcript processing machinery of *Drosophila* (27), and one might consider that it is this which allows the *FLP* gene to be transcribed, the RNA exported from the nucleus, and the message translated. However, the fact that some *FLP* insertions exhibit substantial expression without heat shock argues that these poly-A signals are not a problem for the expression of this gene in *Drosophila*. The evidence concerning expression of FLP in mammalian cells is not clear in this regard, but it is probably prudent to use an *FLP* construct from which the poly-A signals have been removed. This can and has been done without altering the coding sense of the gene (28).

ii. *Promoters for expressing recombinases*

Promoters that have been used to express recombinases in mammalian cells include the metallothionein-I promoter (29), the cytomegalovirus immediate-early promoter, and the Rous sarcoma virus LTR promoter (28, 30). In plant cells the 35S promoter of cauliflower mosaic virus has been used with success (18, 19, 31, 32). A variety of other inducible or developmentally regulated promoters might be appropriately selected by the investigator familiar with his or her chosen organism.

3.2 Mosaics by mitotic recombination

The greatest benefit of using mitotic recombination to generate mosaics is that pre-existing mutations can be analysed. In the case of *Drosophila melanogaster* the collection of mutations available is extremely large—the most recent catalogue lists mutations in close to 4000 genes (33). This mitotic recombination is, in principle, similar to that which can be induced by X-irradiation, but in practice has many distinct advantages. First, the recombination between allelic FRTs is extremely efficient (21). When efficiency was measured as the fraction of heat-shocked animals that became somatic mosaics, nearly all individuals showed mosaicism in the limited portion of the body that was examined. The majority of these animals had certainly experienced multiple independent mitotic recombination events as well as additional recombination in parts of the body that were not scored. FLP-mediated mitotic recombination has also been demonstrated in the male and female germlines. There is normally no meiotic recombination in *Drosophila* males, but a single heat shock was capable of inducing almost 40% germline recombination between two genes flanking allelic *FRT* insertions. A single heat shock can also induce mosaicism in the germlines of virtually all females (34). This extremely high efficiency makes it practical to do experiments that before might have been too difficult or tedious, such as those that require dissection in order to determine whether an appropriate mosaic has been produced. A second advantage is that recombination occurs at known and predetermined sites and all genes distal to that site become homozygous simultaneously. This makes the task of choosing an appropriate marker gene easier. Recombination at the *FRT*s will always make the marker gene and the mutation of interest homozygous at the same time, so the marker may be located anywhere on the chromosome arm as long as it is distal to the *FRT*. A third advantage is that *hsFLP* can be induced with heat shocks that cause little or no cell death, while a typical X-ray dose used to induce mitotic recombination kills about 50% of the imaginal cells in a developing larva.

A minimal set of stocks can be used for mosaic analysis of almost any mutation. A set of five *FRT* insertions in the very proximal portions of each of the major chromosome arms and a few different insertions of *hsFLP* make it a simple matter to generate mosaics for any mutations that lie distal to the sites of *FRT* insertion. In order to accomplish this, the mutation to be examined in mosaics must first be recombined on to the *FRT*-bearing chromosome arm. Second, a marker gene mutation that lies distal to the same *FRT* insertion must also be recombined on to that same chromosome or on to its homologue. The choice of marker gene will depend on several factors:

- the tissue being examined—cuticle markers will be of little use in detecting internal clones
- the process affected by the gene of interest—the marker gene should not interfere with this process
- the developmental stage of mosaic induction—the marker gene should function after the mosaic is generated
- the markers available for the chromosome arm under study

The marker gene and its disposition relative to the mutation of interest (in *cis* or *trans*) should be chosen so that the clone of cells that becomes homozygous for the mutation of interest can be distinguished from both the sister-cell clone and from the parental cell type.

The process of making mosaics by mitotic recombination differs from that of excision (Section 3.3) in several respects: mitosis is absolutely required after the exchange event to make the mosaic; reciprocal daughter cells are generated (thus, a maximum of 50% of cells may be changed to a given genotype); the occurrence of recombination is not a guarantee that mosaicism will follow.

Once recombination has occurred between the *FRT*s on nonsister chromatids, mitosis may then segregate the chromatids that have recombined to opposite poles so that each daughter cell becomes homozygous for the alleles distal to the site of exchange (*Figure 3*). This is the segregation that Curt Stern termed *x* segregation. Recombination events that don't produce mosaics may also occur. Recombination might occur in G1; it would produce one cell with two recombinant chromosomes and an unchanged genotype. Exchange between nonsister chromatids can also fail to produce homozygous daughter cells if it is followed by what Stern called *z* segregation, where the recombinant chromatids segregate to the same pole. Two daughter cells with genotypes identical to the parent cell are produced, although one of these daughters carries recombinant chromosomes (*Figure 3*). In *Drosophila* it seems that *x* segregation exceeds *z* segregation by at least 2:1 (35), but in *S. cerevisiae* the two types of segregation occur equally frequently (36).

The terms *x* and *z* segregation assume that a single crossover has taken place between nonsisters. Because the recombination catalysed by FLP is so frequent, there may be multiple crossovers between sister and nonsister chromatids before the recombinant chromosomes are distributed at mitosis, making it difficult to interpret actual segregations as *x* or *z*. Fortunately these unresolved theoretical considerations do not prevent us from enjoying the benefit of generating mosaics by this route. As a practical matter, mitoses that generate homozygous daughters may be thought of as *x*-type segregations.

The mosaicism that is produced by mitotic recombination does not occur at the time of recombination, but only after the next mitotic division. In some tissues there may be a considerable interval between the time of recombination and the ensuing mitosis. Mitotic recombination induced in abdominal histoblasts of young larvae will not manifest itself as mosaicism until the time of puparium formation when those cells next divide. And, in spite of the earlier discussion on the likelihood that FLP is a short-lived protein, it should not be taken for granted that independent events have occurred simultaneously. Although the frequency of recombination events probably peaks within a few hours after the heat shock, it seems almost certain that, with decreasing likelihood as time goes by, FLP may persist and catalyse recombination in later cell cycles. Some mitotic recombination events may occur at different times as a result of constitutive *hsFLP* activity. And finally, spontaneous mitotic recombination unrelated to FLP, does occur. These spontaneous events are not constrained to occur at the *FRT*s, but could occur at a site between the gene of interest and the gene being used for a marker. Fortunately, spontaneous mitotic recombination is much rarer than FLP-mediated recombination (by a factor of 10^{-4} or less), so in animals where FLP has been induced the vast majority of clones result from recombination at the *FRT*s (but also see Section 4.1).

3.2.1 Perdurance

Proteins or RNAs that are made before the mitotic division that segregates homozygous daughters may persist in these daughter cells and confer on them a wild-type phenotype even though they are now genetically mutant. Perdurance is the term used to refer to this persistence of a gene product in cells that are no longer capable of making that product. Because of the possibility of perdurance it is often best if clones are made at least a couple of cell divisions before they are to be assayed. The degree of perdurance of any gene product will be influenced by the stability of its mRNA and protein, the length of time between genotype segregation and product function, and the number of divisions that have intervened and diluted that product. If the gene has been cloned then the techniques discussed in Section 3.3 may be used to turn the gene on in some cells, rather than turning it off (which is the usual result of mitotic recombination), and eliminate the potential problem of perdurance.

3.2.2 The Minute technique

Minute mutations are a class of dominant mutations that cause a variety of phenotypes, including small bristles, female sterility, reduced viability, recessive lethality, and lengthened development. In an animal that is M/M^+, clones that are M^+/M^+ can be generated. The growth of these M^+/M^+ cells can rapidly outpace that of M/M^+ cells (the M/M cells probably die). By placing a mutation to be analysed on the same arm as the M^+ allele, and heterozygous with the M, mosaics can be generated that have large-sized clones. The growth advantage of M^+/M^+ cells in a M/M^+ animal can also be used to make clones of cells that are homozygous for mutations which are inherently unhealthy and would otherwise not proliferate sufficiently for analysis. A list of suitable *Minute* mutations has been compiled by Lawrence (37).

3.2.3 Choice of markers in *Drosophila*

i. Eye colour markers

Most, but not all, mutations that affect eye colour in *Drosophila* function autonomously, and can usefully serve as markers for mitotic recombination in the eye (34, 38). The *white* gene is an especially good marker because the demarcation between w^+ and w^- clones can be easily seen under the dissecting microscope; even in sections prepared for electron microscopy the w^+ and w^- cells can be distinguished by the presence or absence of pigmented granules (1). Different hypomorphic *white* alleles can be chosen so that both of the reciprocally marked daughter cells may be distinguished from each other and from the parental cells. The use of *white* as a marker is not limited to the marking of clones of X-linked genes. If the X-linked alleles of *white* are nonfunctional, then a hemizygous *white*-bearing transposon can serve as a marker for mitotic recombination. Transposons that carry *white* are available scattered throughout the genome (39). Some transformed copies of *white* have unusual patterns of pigmentation that preclude their use as markers. The chosen transformant should produce pigmentation in order to be useful. A second consideration is whether the *white* gene is driven off its own promoter or a heat-shock promoter as some mini-*white* constructs are. If a heat shock is used to induce FLP synthesis then it is possible that perdurance of the induced gene products could cause pigment to develop in cells that become genotypically *white*$^-$ after mitosis.

The suitability of other eye colour mutations for mosaic analysis can often be improved by placing them in the background of a second eye colour mutation. For instance, *bw*$^-$ clones appear as white patches in the eyes of flies that are also homozygous for a null mutation in either the *vermilion, cinnabar, lightoid,* or *scarlet* genes. Transposons that carry the *brown* gene (40) are now also available and can be used as markers in the same way that *white*-bearing transposons can be used. It should also be noted that the pigments present in wild-type eyes are not evenly distributed by cell type. The *cinnabar* mutation knocks out only one of the two pigment families (xanthomattin) yet the primary pigment cells of cinnabar flies are colourless. The pigment that requires the *brown* gene for its formation is not present in these cells, and so the *brown* gene is not suitable as a mitotic recombination marker for them. If an eye colour mutation other than *white* is chosen as the marker gene it will be necessary to cross out the w^{1118} null allele that many P[>w^{hs}>] stocks carry in favour of a w^+ allele, since *white*$^-$ mutations are epistatic to other eye colour mutations.

Ocelli, Malphigian tubules, and the testis sheath also develop pigmentation that depends on *white* gene function, and, to varying degrees in the different tissues, on other eye colour genes. Lindsley and Zimm (33) can be consulted for details concerning a particular gene or allele. A convenient property of the pigment granules found in all these pigmented tissues is that they can be excited to fluoresce using the standard filter combination designed for FITC on microscopes equipped for epifluorescence. This can greatly aid in distinguishing pigmented from nonpigmented cells.

If P[>w^{hs}>] is to be used as the source of *FRT*s for mitotic recombination, then excision of the *FRT*-flanked w^{hs} gene that will occur along with the mitotic recombination events can confound the interpretation of the resulting mosaic. Thus, chromosomes should be collected that carry only the single *FRT* that remains after w^{hs} excision and these used for the mitotic recombination. Even in a w^+/w^{1118} background, w^{hs} mosaicism can produce a visible variation in the degree of pigmentation and should probably be eliminated even if *white* is not the gene chosen as the marker. The use of a single remnant *FRT* on each chromosome will cause a significant reduction in the frequency of mosaics obtained (Section 3.3.1*ii*), but it should still be frequent enough that a substantial fraction of individuals will be mosaic in any given tissue.

ii. Cuticle markers

The most useful markers for cuticle are those which change only the morphology or colour of a cuticular structure without altering the actual fate of that structure, because in many cases the control of these fates is the matter in question. Discussions of mutations that can be used for marking cuticle may be found in many reference books (4, 37, 41). Most alter bristle or trichome morphology. Suffice it to say that there are a number of possibilities on each chromosome arm. As these marker genes are cloned, and transposons that carry them become available, the choices on each arm will become even greater. Transposons that carry the *yellow* gene are available (42), but over short distances *yellow* function is nonautonomous and it does not mark cuticle in small patches, though it does work well as a bristle marker in small patches. The yellow phenotype is also difficult to score in the presence of mutations that severely affect bristle morphology, such as *singed* and *forked*.

iii. Enzyme markers

The examination of clones in internal tissues is most easily accomplished with stainable enzyme markers. There are null alleles available for some stainable enzymes, in which cases mitotic recombination can be induced in the hetero-zygote to produce *null/null* marked cells (4, 37). The wide availability of transformed copies of *lacZ* and the lack of background staining in wild-type flies will often make it the enzyme of choice. Since most of the available *lacZ* insertions are of the enhancer trap variety it is necessary to select an insertion that expresses in the tissue of interest at the stage that will be assayed. The mitosis that generates the marked cells should also occur early enough so that stainable β-galactosidase activity does not persist in cells that are in fact *lacZ⁻*. The β-galactosidase which is made from enhancer traps typically shows dynamic patterns of staining that suggest the protein decays within a few hours after it is made, so perdurance of β-galactosidase is not likely to be much of a problem.

iv. Antigenic markers

Almost any protein with an available antibody can serve as a marker if a null allele of the gene is available and if it is expressed in the cells of interest. The antigen can be localized with immunofluorescence or chemical localization techniques. Immunofluorescence with antibodies against β-galactosidase (or other enzymes) can also be used as an alternative to enzyme activity stains if the dark stain would obscure the cytology of the tissue.

v. Female germline

One especially useful method for making female germline clones uses dominant female sterile (DFS) mutations (4, 34). A chromosome arm that carries a DFS mutation is made heterozygous with an arm carrying the mutation that one wishes to make homozygous in the female germline. Cells that become homozygous for this mutation are at the same time relieved of the effect of the DFS mutation. The DFS mutation most used for this purpose is the X-linked *ovo^{D1}* mutation. This allele acts specifically in the germline and prevents the production of any eggs. Thus, if eggs are laid by an *ovo^{D1}/ovo^+* mother, they will have resulted from a mitotic recombination event that made a germline clone homozygous for *ovo^+* and all other alleles on that arm distal to the site of exchange. It then becomes quite easy to determine whether a recessive female sterile mutation acts in the germline (in which case the females will remain sterile) and to assess germline effects of recessive lethal mutations.

3.2.4 Mitotic recombination in other organisms

One reason that FLP-mediated mitotic recombination works so efficiently in *Drosophila* is undoubtedly because homologous chromosomes pair with each other not only in meiosis but in somatic and pre-meiotic germ cells. This pairing probably acts to maintain the allelic *FRT*s in close association with one another so that they can be readily recombined when FLP is synthesized. It is not clear whether the homologous chromosomes of other organisms maintain such an intimate association (43). Achieving high frequencies of mitotic recombination in other organisms will probably depend on a certain degree of pairing of homologues. There has been one report of a possible instance of mitotic recombination in the mouse, but the crucial evidence for mitotic recombination, the production of twin spots, was lacking (44). Until methods of inducing site-specific recombination in intact individuals are produced for other organisms with sufficiently developed genetics this will remain a matter for speculation.

3.3 Mosaics of cloned genes

An asset of using site-specific recombination to make mosaics is that the small, defined recombination targets allow one to build constructs that can generate very precisely defined and controlled mosaicism, and this mosaicism can be produced in organisms that may not be susceptible to the induction of mitotic recombination. The following discussion is meant to summarize some of the ways that this asset has been exploited and to suggest some additional possibilities. It is by no means an exhaustive discussion of all possible variations.

The first requirement for generating an informative mosaic is that a genetic background be available that will allow the gene that is being manipulated to have a dominant effect. This can be obtained if recessive mutations in the gene of interest are available. In this case the effect of a gene construct that can provide wild-type function can be examined in a homozygote for the recessive mutation, where it will act as a dominant allele. If the clone of an allele that is dominant to wild-type is available or can be constructed *in vitro* then these may be used even when mutant strains are not available.

Although the examples discussed thus far have made use of induced loss-of-function mosaicism, site-specific recombination allows one to just as easily produce gain-of-function mosaics. This is especially advantageous when mosaics are used to define the time at which a gene's expression is required for function. By determining the latest stage at which a gene can be turned on and fulfil its role in development and the earliest stage at which it can be turned off and yet provide full function, it is possible to define the time at which that gene's expression is developmentally significant. The fact that recombination frequencies can be made to approach 100% (excisions/cell) greatly reduces the observations needed for this type of experiment. Linking a gene to a heat-shock promoter to provide transient expression can also address this question, but the tissue and cell-type specificity conferred by a gene's own promoter will be lost, and if expression is required over a lengthy period or at precise levels then heat-shock-promoted expression may prove cumbersome or ineffective.

3.3.1 Factors that influence recombination frequency

i. FRT distance

The single factor that most strongly influences the frequency of FLP-mediated recombination between *FRT*s, after the heat-shock conditions, is the distance between those *FRT*s. When FLP is used *in vitro*, recombination frequency decreases markedly as the separation between *FRT*s increases from approximately 400 bp to 3400 bp (45). Based on our estimates from somatic mosaicism, the frequency of recombination between *FRT*s may drop as much as two-fold when the distance between them increases from approximately 5 to 18 kb. *FRT*s that are separated by an estimated few hundred kilobases recombine at a frequency of a few per cent—an additional ten-fold decrease. Recombination between FRTs located on heterologous chromosomes is quite rare, occurring in perhaps one in 10 000 cells after *hsFLP* induction (unpublished data). Because of this decreased recombination frequency with distance one may include in the animals to be made mosaic more than one insertion of *FRT*s and be confident that recombination between them (that might generate chromosomal rearrangements) will be so infrequent that it need not be a matter of concern. It should be emphasized that these frequencies are rough estimates arrived at in different ways using different constructs. They also do not take into account possible context effects; however, in other experiments the variation in recombination frequency between different insertions of the same construct was almost always less than a two-fold difference.

ii. FRT copy number

Another source of variation in recombination frequency is the number of *FRT* copies present (21). When the mosaics produced by mitotic recombination were scored while varying the number of *FRT*s at the allelic sites from 2/2 to 2/1 to 1/1, the frequencies of mosaic animals fell from 97% to 37% to 1%.

This assay was performed using the high constitutive level of activity in the notum provided by the *hsFLP*1 insertion. Higher levels of recombination could be induced by using a heat shock—the frequencies of mosaicism produced were then 99.5%, 59%, and 44%, respectively. Thus, variation in *FRT* dosage can effectively alter the frequency of recombination. One experimental use of this can be to reduce the frequency of spontaneous mosaics that arise from constitutive FLP activity.

Although I have called this an effect of *FRT* copy number, it may actually be due, in whole or part, to the loss of the w^{hs} gene from the region and a resulting change in chromatin configuration. The experiments to differentiate these possibilities have not been done, but if one uses an insertion of $P[>w^{hs}>]$ as the source of *FRT*s then these relationships should hold. This demonstration of copy number effect relied on recombination between *FRT*s on homologous chromosomes; a similar relationship presumably applies to intrachromosomal recombination.

iii. FRT size

The minimal complete *FRT* is considered to be 34 bp. This is readily recognized by FLP when integrated in *Drosophila* chromosomes (R. Holmgren, personal communication) and on extrachromosomal circles in mosquito embryos (17). A slightly larger *FRT* carrying the third 13 bp repeat is efficiently recognized by FLP in mammalian cells. It is not known whether the third repeat is required in these cells. There is some evidence that the third repeat may enhance the efficiency of recombination. I have used a 200 bp fragment from the 2μ plasmid taken from about the XbaI site in the spacer region of the *FRT* (unpublished data) and the entire 599 bp repeat from the 2μ plasmid *in vivo* (20). Both are efficiently utilized by FLP. The 599 bp repeat may have a slightly increased efficiency of recombination when compared with the 34 bp *FRT*, but if so it is not a large effect. Notwithstanding that different constructs, insertion sites, and methods of assay have been used with each *FRT*, the 34 bp *FRT* (and perhaps the 28 bp minimal *FRT*) can certainly be used for efficient recombination, in agreement with results obtained *in vitro*.

3.3.2 *FRT* placement

i. Outside gene

This configuration can be easily utilized to generate mosaics in which all cells initially carry the functional gene and then some fraction lose that gene (*Figure 4A*). In order that this method results in loss of function the gene must be excised prior to the last round of mitosis.

Mosaicism and cell division

When FLP is used to excise the w^{hs} gene of $P[>w^{hs}>]$ (shown in *Figure 5A*) by heat-shocking late third instar larvae or early pupae the mosaics that are generated exhibit an apparently precise correlation between the pattern of

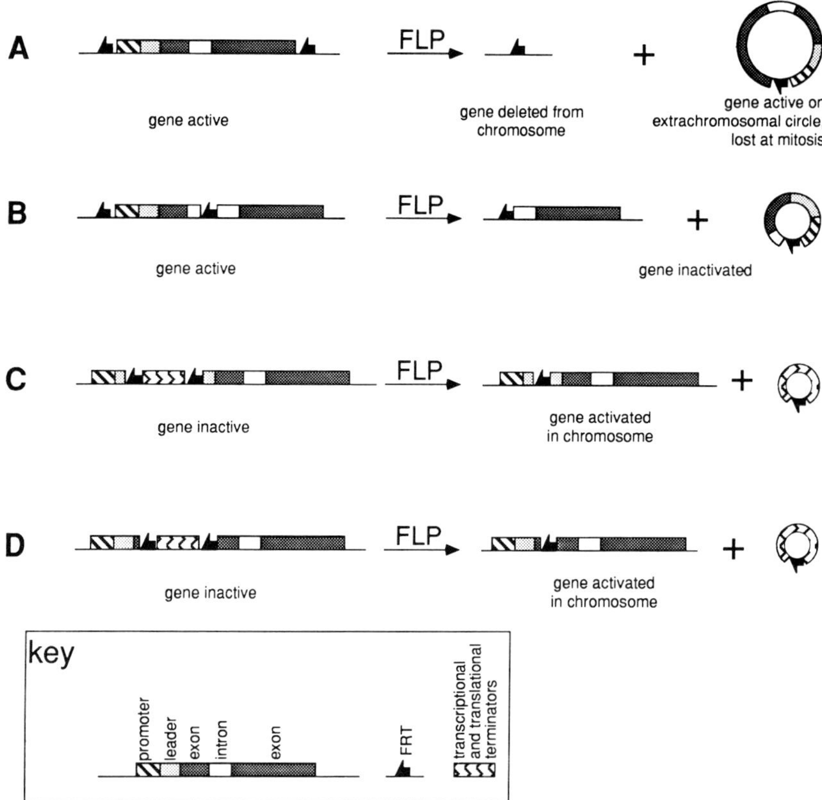

Figure 4. The effects of *FRT* placement. A gene can be inactivated by FLP: (A) *FRT*s flanking a gene; (B) One *FRT* in an intron and one *FRT* outside the gene. A gene can be activated by FLP: (C) An *FRT*-flanked termination cassette placed in the leader sequence of a gene: (D) An *FRT*-flanked termination cassette placed within the protein-coding region of a gene. See text for further details.

w^{hs} loss and the known pattern of DNA replication (and cell division) in the eye imaginal disc at this stage (1, 20). Those portions of the eye disc that show a loss of *white* function (*Figure 5A*) are arranged in a pattern very similar to that of cells undergoing replication (as visualized by [^{3}H]thymidine incorporation). This correlation was interpreted as indicating that FLP functioned only in actively dividing cells but we now know this is not true. A *white* gene was constructed with an *FRT* within the first intron and a second *FRT* downstream of the gene in inverted orientation. FLP acts on this gene to invert a portion of it and make it nonfunctional. This inversion can be made to occur in the germline and progeny can be collected that carry the gene in the 'off' configuration. If flies that carry this turned-off gene and *hsFLP* are heat-shocked the gene will be turned back on and pigmentation will be restored to a

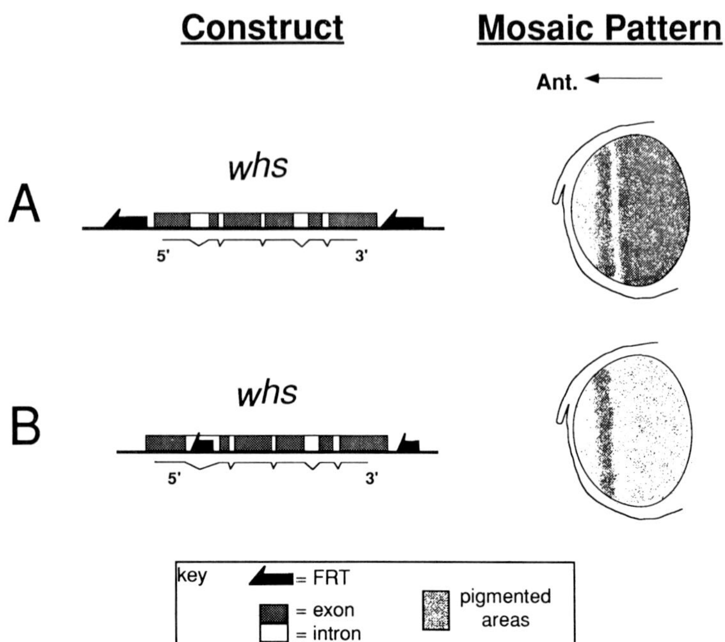

Figure 5. Eye-colour mosaicism resulting from *hsFLP* induction in late third instar. The constructs and the mosaicism they produced are shown. See text for details. Ant. = anterior.

fraction of the cells in the eye. Pigmentation can be restored even if the flies are not heat-shocked until after they eclose as adults—many days after the last mitoses in the eye (unpublished data). Thus, it is clear that FLP can mediate recombination in nondividing cells.

What then can explain the pattern of mosaicism shown in *Figure 5A*? First, we suppose that in the nondividing cells in the posterior of the eye (the posterior pigmented portion) the w^{hs} gene is excised (see Section 3.3.2*ii* below) but remains intact on an extrachromosomal circle and is expressed. Second, in most of the cells that will undergo further division (the region in front of the posterior pigmented portion), the w^{hs} gene is excised and then lost from one or both daughters at mitosis. Third, the cells in the more anterior pigmented stripe appear to respond to heat shock at a lower level than cells in the rest of the eye (Section 3.3.1*iii*). Little FLP is made in these cells and loss of w^{hs} is infrequent. The heat-shock response is not entirely eliminated in these cells, as more severe heat shocks do in fact diminish the pigmentation of the anterior stripe.

Two lessons can be drawn from these results: first, although FLP does not require cell division in order to function, some types of mosaicism generated by FLP do require cell division to manifest themselves; and second, cells in different stages of development are not always equally responsive to heat shock.

ii. In an intron

As mentioned in the previous section, we have placed a 200 bp *FRT*, consisting of approximately 100 bp to each side of the central XbaI site, within the first intron of the w^{hs} gene (as in *Figure 4B*). Expression of w^{hs} is not affected by the presence of this sequence in the intron. A second directly repeated *FRT* was placed upstream of the gene in one construct and downstream in another. Recombination between the *FRT*s in both constructs was efficient and could be used to inactivate the gene even after the final mitotic divisions in the eye because this recombination event cuts the gene in two and neither product is functional (the mosaicism generated is shown in *Figure 5B*). If this method is used to inactivate a gene it would be wise to make sure that the portion of the gene that remains connected to the promoter is incapable of providing or interfering with function.

When inverted repeats are used to invert a portion of a gene, thus inactivating it, high levels of FLP tend to drive the orientation of such a gene to equilibrium—that is, half the cells have an inactive gene, and half an active gene. When using excisional inactivation high levels of FLP tend to produce 100% inactivation.

iii. In leader sequence

A 34 bp *FRT* inserted between the promoter of a gene and the translational start can be compatible with that gene's function. One use of this placement has been to arrange for a marker gene to be turned on by a site-specific recombination event (*Figure 4C*). The *lacZ* gene was placed under the control of an *Actin5C* promoter; between the promoter and the *lacZ* gene a cassette was inserted that consisted of two directly repeated 34 bp *FRT*s flanking the transcriptional and translational stop regions of the *hsp70* gene. This prevented *lacZ* expression until FLP catalysed the excision of the termination sequences (R. Holmgren, personal communiction). If high levels of FLP were induced so that excision approached 100% β-galactosidase was then produced with a fairly ubiquitous pattern of expression. If conditions are used so that excision is rare, then such a construct can be used to produce clones of marked cells that are distinguished as blue-staining cells on a white background, a result that could not be readily obtained before the advent of site-specific recombination. Placement of a cassette like this may be critical. If such a cassette were placed in an intron, and if splicing occurs cotranscriptionally, the transcriptional stop might be spliced from the growing message before it could effect termination. If this arrangement prevents *lacZ* expression because it blocks translation, then of course it also would not work if placed in an intron.

Similar placements of recombination sites for other recombinases have been successfully used in mammalian cells and in plant cells (18, 19, 31, 46).

iv. In coding sequence

The 34 bp complete *FRT* has open reading frames on both strands (see *Figure 1*). This has allowed for the placement of an *FRT* within the coding sequence of a gene (*Figure 4D*). A *neomycin-resistance* gene flanked by two *FRTs* was inserted into the *lacZ* coding sequence so that upon excision the remaining *FRT* would be in frame with *lacZ*. When transfected into mammalian cells this *lacZ* gene was inactive, but upon excision of the *neo* gene by FLP *lacZ* was strongly expressed (28). Although β-galactosidase function is relatively immune to disruption by protein fusions, the function of some proteins may not be compatible with the addition of the few amino acids encoded by the *FRT*, and for those proteins this method would not be appropriate. If such a placement of *FRTs* is used it may be wise to put it very near the start of translation so that no partially functional peptide is produced prior to FLP-mediated excision of the *FRT*-flanked cassette.

v. Nested genes

Genes can easily be nested with these *FRT* placement techniques to provide mosaicism for a gene of interest and a simultaneous marking of cells. For instance, a cell that loses an *FRT*-flanked gene (such as the *neo* gene in the previous example) will become marked by activation of *lacZ*. Remember that an excised gene is not lost until the next mitosis, and even then it is possible that the extrachromosomal gene may be passively retained by one of the daughter cells. To ensure that the pattern of β-galactosidase staining coincides precisely with loss of function, it is recommended that the excision event be induced several cell divisions before the time that the gene in question functions.

3.3.3 Additional considerations

i. Unequal sister-chromatid exchange: direct repeats

FLP occasionally catalyses exchange between staggered *FRTs* on sister chromatids. This unequal sister-chromatid exchange (USCE) results in the production of two tandem copies of the region flanked by *FRTs* (*Figure 6*). Recovery of this amplified arrangement is infrequent at high levels of FLP—this condition tends to produce predominantly excision. At lower levels of FLP the amplified product is recovered in perhaps 2–4% of cells when the *FRTs* are 5–6 kb apart. These amplification events may be of concern if the biological system in question is sensitive to this sort of variation in gene dosage.

ii. Unequal sister-chromatid exchange: inverted repeats

When *FRTs* are arranged as inverted repeats the consequences of USCE between them is the generation of dicentric chromosomes and acentric frag-

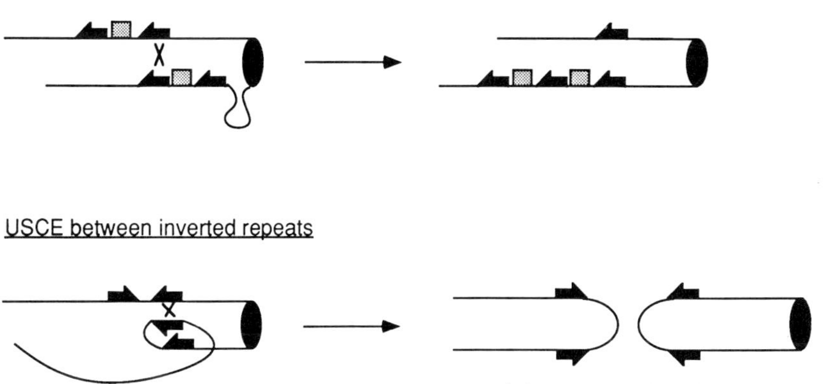

Figure 6. The consequences of unequal sister chromatid exchange (USCE) between direct and inverted repeats of *FRT*s. The large solid oval indicates the centromere.

ments (*Figure 6*) (47). If the site where the inverted *FRT*s have integrated is sufficiently proximal on an arm (say 1/4 or so of an arm) then the frequency of USCE exchange events can be easily measured by metaphase cytology of larval neuroblasts (48) 6–8 hours after *hsFLP* induction. The results of an USCE event are easily visualized as the distal acentric portion of the chromosome floating free of the proximal dicentric portion. The fraction of metaphases with acentric/dicentric cytology can then be scored. We have in some cases observed close to 90% of the nuclei in such spreads exhibiting this characteristic cytological phenotype. These are not happy flies, although many do survive. If the formation of these dicentrics is induced in late third instar or the early pupal stage, the adults that do eclose show a number of characteristic phenotypes, including roughened eyes, scalloped wings, and etched tergites. If the inverted *FRT*s lie far enough apart then this USCE may be reduced in frequency to such an extent that it does not cause problems. However, unless the experimental approach demands the use of inverted *FRT*s, it is probably best to design a scheme that makes use of directly repeated *FRT*s.

The frequency of USCE products obtained with inverted repeats is much higher than with direct repeats. This difference probably reflects the facts that when the acentric and dicentric fragments are generated they are free to diffuse away from each other, effectively preventing recombination that might restore the chromosome to normal. Also, additional intrachromosomal recombination events between direct repeats can excise amplified genes, but intrachromosomal events following the generation of dicentric chromosomes do not restore the full length chromosome.

4. Potential drawbacks

As with any new technique, uncharacterized aspects have the potential to produce unexpected problems. Some cautions have already been discussed in the sections above. Below are a few more. I emphasize that these are only potential problems; there is little evidence that they are, in fact, problems in experimental situations.

4.1 Cryptic sites

A valuable aspect of site-specific recombination is that the recombination occurs at determined sites. If a genome harbours sequences that can act as substrates for recombinase it would detract somewhat from the use of that recombinase for generating mosaics. (Of course cryptic sites might have their own uses.) The expected chance occurrence of the 28 bp minimal *FRT* is approximately once in 10^{17} bases—not very likely. Yet since many base changes within an *FRT* are tolerated, albeit with reduced efficiency, there may be sequences in a genome that *FLP* can recognize and recombine at a low frequency. In fact the *S. cerevisiae* genome does have sequences within it that can serve as very inefficient substrates for Cre recombinase (49). We have seen no evidence that there are cryptic *FRT* sites in the genome of *D. melanogaster*, but we have not examined all chromosome arms nor looked in ways that might detect sites that are used with greatly reduced efficiency.

A search of databases for *D. melanogaster* sequences with similarity to the *FRT* uncovered a few surprisingly good matches: one in the G element found in rDNA, one in the *Abl* gene, and one in the *l(2)gl* gene. The effects of multiple base substitutions in the 13 bp repeats of the *FRT* have not been characterized, but if assumptions are made that each change acts independently to reduce recombination efficiency and that alterations in the spacer do not matter at all, then one can predict how well an *FRT*-like sequence would work as a substrate for FLP. In the cases mentioned, the predicted efficiencies are approximately 2000, 2000, and 5000-fold less than a wild-type *FRT*, respectively. The ability of the *Abl* sequence to serve as a substrate for FLP is probably reduced further than this because its spacer sequence would be only 6 bp long.

4.2 Aberrant recombination

It is possible that at some low level, FLP-mediated recombination may proceed abnormally and be resolved in unpredictable ways by FLP or by cellular DNA repair mechanisms. Aberrant products have been noted when purified FLP is used for *in vitro* reactions (50), and a small number of unexpected products have been recovered in one experiment *in vivo* (17). We have seen little to suggest that aberrant events occur with appreciable frequency in *Drosophila*.

One can also conceive of a normal recombination event with aberrant effects; if an *FRT* that has already replicated recombines with an *FRT* on the homologue that has not yet replicated chromosome aberrations will result if not resolved by further recombination. Again, if such events do occur they must be rare.

4.3 FLP temperature dependence

Pan *et al.* (51) have found that FLP synthesized in *E. coli* growing at 37°C shows 40-fold less activity than FLP synthesized in *E. coli* growing at 25°C. Thus, incubating organisms at 25°C after inducing an *FLP* gene, a condition that always attains in *Drosophila*, may increase the frequency of recombination. However, FLP has worked quite efficiently in mammalian cells grown at 37°C (28).

5. Appendix

5.1 *Drosophila* stocks

Stocks of *Drosophila melanogaster* that carry *hsFLP, 70FLP*, and P[$>w^{hs}>$] may be obtained from the Indian Drosophila Stock Center, Dept of Biology, Indiana University, Bloomington, IN 47401, USA.

5.2 *Drosophila* methods

Additional information regarding laboratory methods with *Drosophila* may be found in books by Roberts (52) and Ashburner (48). Information on mutations has been compiled by Lindsley and Zimm (33).

5.3 Original references

Original references to much of the work on site-specific recombination may be found in the reviews by Cox (11, 12) and Hoess and Abremski (13). References to the original papers on making mosaics in *Drosophila* may be found in Ashburner's book (4).

Acknowledgements

I thank Bob Holmgren for allowing me to discuss his unpublished work, and numerous other colleagues for stimulating discussions and ideas concerning the use of site-specific recombination. Kelly Gordon and Bill Sullian provided helpful comments on the manuscript. Work in our laboratory is supported by Grants JFRA-370 and IRG-178B from the American Cancer Society, and Grant HD28694 from the National Institutes of Health.

Note added in proof

Lakso *et al.* (*Proceedings of the National Academy of Sciences USA*, **89**, 6232) have recently demonstrated the use of *Cre-lox* site-specific recombination in transgenic mice.

References

1. Ready, D. F., Hanson, T. E., and Benzer, S. (1976). *Dev. Biol.*, **53**, 217.
2. Rubin, G. M. (1991). *Trends in Genet.*, **7**, 372.
3. Greenwald, I. (1989). *Trends in Genet.*, **5**, 237.
4. Ashburner, M. (1989). Drosophila: *A laboratory handbook*. Cold Spring Harbor Press, Cold Spring Harbor, NY.
5. Herman, R. K. (1984). *Genetics*, **108**, 165.
6. LeDouarin, N. and McLaren, A. (ed.) (1984). *Chimeras in developmental biology.* Academic Press, London.
7. Kimmel, C. B. and Warga, R. M. (1988). *Trends in Genetics*, **4**, 68.
8. Poethig, S. (1989). *Trends in Genetics*, **5**, 273.
9. Huszar, D., Sharpe, A., Hashmi, S., Bouchard, B., Houghton, A., and Jaenisch, R. (1991). *Development*, **113**, 653.
10. Craig, N. L. (1988). *Annu. Rev. Genet.*, **22**, 77.
11. Cox, M. M. (1988). In *Genetic recombination* (ed. R. Kucherlapati and G. R. Smith), pp. 429–443. American Society for Microbiology, Washington, DC.
12. Cox, M. M. (1988). In *Mobile DNA* (ed. D. E. Berg and M. M. Howe), pp. 661–668. American Society for Microbiology, Washington, DC.
13. Hoess, R. H. and Abremski, K. (1990). In *Nucleic acids and molecular biology* (ed. F. Eckstein and D. M. J. Lilley), Vol. 4, pp. 99–109. Springer-Verlag, Berlin.
14. Petersen, R. B. and Lindquist, S. (1989). *Cell Regulation*, **1**, 135.
15. Amin, A. A. and Sadowski, P. D. (1989). *Mol. Cellular. Biol.*, **9**, 1987.
16. Huang, L. C., Wood, E. A., and Cox, M. M. (1991). *Nucl. Acids Res.*, **19**, 443.
17. Morris, A. C., Schaub, T. L., and James, A. A. (1991). *Nucl. Acids Res.*, **19**, 5895.
18. Onouchi, H., Yokoi, K., Machida, C., Matsuzaki, H., Oshima, Y., Matsuoka, K., Nakamura, K., and Machida, Y. (1991). *Nucl. Acids Res.*, **19**, 6373–6378.
19. Maser, S. and Kahmann, R. (1991). *Mol. Gen. Genet.*, **230**, 170.
20. Golic, K. G. and Lindquist, S. (1989). *Cell*, **59**, 499.
21. Golic, K. G. (1991). *Science*, **252**, 958.
22. DiDomenico, B. J., Bugaisky, G. E., and Lindquist, S. (1982). *Cell*, **31**, 593.
23. Monsma, S. A., Ard, R., Lis, J. T., and Wolfner, M. F. (1988). *J. Exp. Zoology*, **247**, 279.
24. Zimmerman, J. L. and Cohill, P. R. (1991). *The New Biologist*, **3**, 641.
25. Bonner, J. J., Parks, C., Parker-Thornburg, J., Mortin, M. A., and Pelham, H. R. B. (1984). *Cell*, **37**, 979.
26. Lindquist, S. (1986). *Annu. Rev. Biochem.*, **55**, 1151.
27. Yost, H. J. and Lindquist, S. (1986). *Cell*, **45**, 185.
28. O'Gorman, S., Fox, D. T., and Wahl, G. M. (1991). *Science*, **251**, 1351.

29. Sauer, B. and Henderson, N. (1988). *Proceedings of the National Academy of Sciences USA,* **85,** 5166.
30. Sauer, B. and Henderson, N. (1990). *The New Biologist,* **2,** 441.
31. Odell, J., Caimi, P., Sauer, B., and Russell, S. (1990). *Mol. Gen. Genet.,* **223,** 369.
32. Dale, E. C. and Ow, D. W. (1990). *Gene,* **91,** 79.
33. Lindsley, D. L. and Zimm, G. G. (1992). *The genome of* Drosophila melanogaster. Academic Press, San Diego, California.
34. Chou, T. and Perrimon, N. (1992). *Genetics,* **131,** 643.
35. Pimpinelli, S. and Ripoll, P. (1986). *Proceedings of the National Academy of Sciences USA,* **83,** 3900.
36. Chua, P. and Jinks-Robertson, S. (1991). *Genetics,* **129,** 359.
37. Lawrence, P. A., Johnston, P., and Morata, G. (1986). In Drosophila: *a practical approach* (ed. D. B. Roberts). IRL Press, Oxford.
38. Phillips, J. P. and Forrest, H. S. (1980). In *The genetics and biology of* Drosophila (ed. M. Ashburner and T. R. F. Wright), Vol. 2d, pp. 541–623. Academic Press, London.
39. Hazelrigg, T., Levis, R., and Rubin, G. M. (1984). *Cell,* **36,** 469.
40. Dreesen, T. D., Henikoff, S., and Loughney, K. (1991). *Genes and Dev.,* **5,** 331.
41. Postlethwait, J. H. (1978). In *The genetics and biology of* Drosophila (ed. M. Ashburner and T. R. F. Wright), Vol. 2c, pp. 359–441. Academic Press, London.
42. Chia, W., Howes, G., Martin, M., Meng, Y. B., Moses, K., and Tsubota, S. (1986). *EMBO J.,* **5,** 3597.
43. Tartof, K. D. and Henikoff, S. (1991). *Cell,* **65,** 201.
44. Panthier, J., Guenet, J., Condamine, H., and Jacob, F. (1990). *Genetics,* **125,** 175.
45. Gronastajski, R. M. and Sadowski, P. D. (1985). *J. Biol. Chem.,* **260,** 12 328.
46. Sauer, B. and Henderson, N. (1989). *Nucl. Acids Res.,* **17,** 147.
47. Falco, S. C., Li, Y., Broach, J. R., and Botstein, D. (1982). *Cell,* **29,** 573.
48. Ashburner, M. (1989). Drosophila: *A laboratory manual.* Cold Spring Harbor Press, Cold Spring Harbor, NY.
49. Sauer, B. (1992). *J. Mol. Biol.,* **223,** 911.
50. Meyer-Leon, L., Huang, L.-C., Umlauf, S. W., Cox, M. M., and Inman, R. B. (1988). *Mol. Cell. Biol.,* **8,** 3784.
51. Pan, H., Clary, D., and Sadowski, P. D. (1991). *J. Biol. Chem.,* **266,** 11 347.
52. Roberts, D. B. (ed.) (1986). Drosophila: *a practical approach.* IRL Press, Oxford.

2

Cell transplantation

ANDREAS PROKOP and GERHARD M. TECHNAU

1. Introduction

1.1 Why transplant cells?

Cell transplantation is a powerful tool for studying cellular interactions during development. It allows the behaviour of cells of known origin in various kinds of environments to be studied. Thus, cell transplantation has been used in both vertebrates and invertebrates to address several types of questions. For example, isotopic and isochronic transplantations of cells have been used to study the developmental potential of cells (1–9). Heterotopic transplantations allow one to test for the regional determination of cells and their ability to communicate with the surrounding tissue (1, 10–12). Heterochronic transplantations allow one to test for the reversibility of the determination of the transplanted cells or for the time period in which particular cellular communication processes are active (13, 14). Using heterogenetic transplantations (transplantation of mutant or transformed cells into wild-type hosts or vice versa) it is possible to test for the interrelations between genetic functions and for the autonomy or nonautonomy of phenotypes (15, 16). Interspecific transplantations of cells have been used to study the functional compatibility of mechanisms that involve intercellular communication (17, 18).

In this chapter we describe the method for single cell transplantation in *Drosophila* embryos. The chapter describes the instrumental set-up, discusses various alternatives for experimental design, describes the technique of transplantation, and details different ways of processing and analysing preparations.

1.2 Overview

The following steps are involved in cell transplantation:

(a) Design an experimental strategy.

(b) Prepare the instrument set-up, including the microscope, micromanipulator, equipment for video microscopy (if necessary), and the needles.

(c) Prepare the donor and host embryos. Dechorionate, orientate, fix to a slide, desiccate, and cover them with oil.

(d) Inject markers into suitable donors.

(e) Transplant (single) cells into suitable unlabelled hosts.

(f) Perform *in vivo* video microscopy during embryogenesis (optional).

(g) Dissect and fix the hosts at any developmental stage you are interested in and histochemical stain for your markers.

(h) Embed them for analysis.

2. Instrument set-up

As there is a variety of instrument set-ups that can be used for transplantation experiments (19–21, 34) we will concentrate on general advice.

2.1 The microscope

For a compound microscope to be suitable for transplantation there are several requirements:

(a) Nomarski optics are necessary for visualizing single cells inside the capillary.

(b) If you are intending to use *in vivo* controls or to trace the development of fluorescently labelled cells by video documentation, the compound microscope has to be equipped with the respective filters and light source for epifluorescence and a c-mount tube for the camera.

(c) The compound microscope can be either a standard or an inverted one.
   i. A standard microscope allows only for a rather small working distance between the lens and the embryo. Thus the objective lens will not be a higher magnification than about 20 ×. In combination with 15 × or 20 × eyepieces the total magnification will be about 300 × or 400 ×. This is sufficient for single cell transplantation. The microscope's stage should be fixed, so that the focusing is executed by moving the tube instead of the stage (thus, the transplantation needle and the embryos will stay at the same level while you focus).
   ii. Inverted microscopes are recommended, because they allow work with lenses up to a magnification of 100 ×. A good working magnification is provided by a 40 × objective lens in combination with 12.5 × eyepieces.

2.2 The needles

Two types of needles are used: needles for the injection of markers and needles for transplantation.

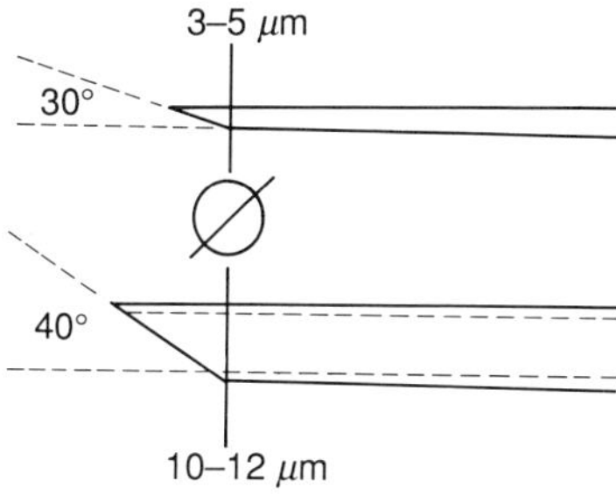

Figure 1. Size and form of capillary tips suitable for injection of marker solutions (top) and for transplantation (bottom). The angles of the bevels and the outer diameters at the distalmost capillary tips are indicated. The diameter of the transplantation capillary should not increase significantly over a distance of 100–200 μm.

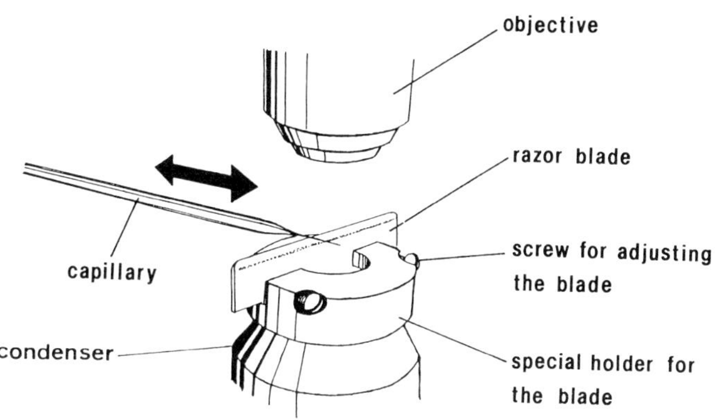

Figure 2. Set-up for cutting transplantation capillaries designed by K. Lüer (*Protocol 1*). The capillary is fixed to the microscope stage and can be moved forwards and backwards until its diameter is about 10–12 μm at the point of connection with the razor blade. The razor blade is fixed with a special holder on top of the condenser.

Protocol 1. Preparation of injection- and transplantation-needles

1. Take soda-lime glass capillaries (outer diameter 1 mm, wall 0.1 mm) and pull them on a commercial pipette puller that allows both the heat and pulling strength to be varied. A very simple puller, which can be manufactured by your workshop, was designed by Zalokar (22).

2. Change the parameters of the puller until you get ideal needles (see *Figure 1*). Some pullers can produce needles that are long (with only a slight change of diameter over a long distance) and have a narrow tip, too. These can be used for injection as well as for transplantation (if you cut off the tip

Protocol 1. *Continued*

of the needle at the position with the right diameter; see below). Other
pullers require different parameters to produce needles for either injection
or for transplantation.

3. Cut transplantation needles with fine forceps, under a dissecting microscope, at a position where the outer diameter is about 10–12 μm. Control
   the result under a compound microscope equipped with a micrometer
   scale. To get reproducible results K. Lüer, in our laboratory, established
   the following method:

   (a) Fix a razor blade with the help of a special holder on to the condenser
       of your compound microscope. This holder carries a hole in the
       middle that allows you to look through it (*Figure 2*).
   (b) Fix a capillary with some plasticine on to the microscope's stage
       (perpendicular to the razor blade) and get its tip into focus.
   (c) Move the condenser up until the blade is focused and touches the tip
       of the needle.
   (d) Move the microscope's stage with the needle perpendicular to the
       blade until the diameter at the point of contact is about 10–12 μm (use
       a μm-scale inside the eyepiece that coincides with the position of the
       razor blade).
   (e) Stroke the needle with the tip of a second pulled capillary along
       the blade until the transplantation capillary breaks at the point of
       contact.

4. Grind the needles on a commercial capillary tip grinder.

   (a) Grind injection needles with an angle of about 20–30° (*Figure 1*), they
       should be sharp enough to penetrate the embryos easily. Usually it
       is sufficient to grind them for less than a minute to obtain a tip with
       an outer diameter of 3–5 μm. If the needle, however, is too thin to
       suck up marker solution easily, it is necessary either to grind it some-
       what longer or to snap off the distalmost tip with forceps before
       grinding.
   (b) Grind transplantation needles for about 1–2 minutes, as the broken
       glass at their tips might disrupt cells during the transplantation pro-
       cess. The bevel should be about 35–40° (sharp enough to penetrate
       the embryo but blunt enough to place the donor cells precisely into the
       host's epithelium; *Figure 1*).

5. Mark the posterior end of the capillary with a pen while it is in the grinder
   holder. This will help you to get the bevel to the right orientation when
   you fix the capillary to the micromanipulator.

6. Clean the tip of the capillary. Suck water in and out followed by acetone,
   with the help of a hose and a single-use syringe (see Section 2.3). Used

capillaries can be cleaned with chromosulfuric acid and subsequently with water and acetone.

7. Store the empty capillaries in a dust-free chamber fixed on plasticine.

2.3 Set-up for micromanipulation

(a) Stick the posterior end of the capillary into a polyethylene hose. Polyethylene does not collapse while sucking. Prepare the hose as follows:
- Take a 10 cm piece of 6 mm (outer diameter) transparent polyethylene tubing, heat it over a small flame and pull it slowly to a length of about 50 cm.
- Cut the ends of the pulled section so that the diameter fits the capillary on one end and the nozzle of a single-use syringe on the other.

(b) Fix the air-filled capillary to the holder of a micromanipulator (use the mark at the end of the needle to orientate so that you will see the bevel under the microscope). We use the manual micromanipulator from Leitz which is equipped with a very precise mechanism and which efficiently prevents vibration of the capillary.

(c) Injection and suction is achieved with an air-filled single-use syringe. This system is easier to handle than oil-filled systems.

3. Choice of cell markers

There are many ways of labelling cells for transplantation experiments. In the following we describe several possibilities.

3.1 Genetic markers

Genetic markers are mutations or alleles of genes, the phenotypic expression of which can easily be distinguished and analysed structurally or histochemically. They should be expressed cell autonomously and ubiquitously in those tissues you are interested in.

3.1.1 External genetic markers

These markers affect the colour of the epidermis (for example *ebony, yellow*) or the number or shape of cuticle structures, such as hairs and bristles (for example *multiple wing hairs, forked*). Most markers are not expressed before the adult stage, but there are markers available for larval and even for embryonic stages. For lists and descriptions of external markers see references 21 and 23.

3.1.2 Internal genetic markers

Markers for internal organs are normally (housekeeping) genes coding for enzymes that are more or less ubiquitously expressed and that can be detected histochemically (for example alcohol dehydrogenase, succinate dehydrogenase; for further internal markers see references 21 and 23). Such donors have to be combined with hosts deficient for that enzymatic function. There are also useful strains available with a gain of enzymatic functions (for example aldehyde dehydrogenase (24); β-galactosidase (25–27)). β-galactosidase (β-gal) especially has become a powerful tool:

(a) The enhancer trap technique (27–30) has led to a mass generation of β-gal expressing strains, so that markers for all tissues are available.

(b) According to the type of β-gal construct the label is either restricted to the nucleus (suitable for cell counts in transplants) or marks the cytoplasm (suitable for identification of cell types).

(c) Many staining techniques for β-gal have been established allowing combinations with several other markers, e.g. antibody (31).

(d) As a host you can use a viable *Drosophila* strain lacking endogenous β-gal (32).

3.1.3 Experimental design

Take a donor fly strain which ubiquitously expresses the internal marker, at least in that tissue you are interested in. Among external markers, take the most conspicuous available for donors (as you will have to detect small epidermal patches). If you have the choice of taking either strain as donor or host, take the more viable strain as host (it will result in better yields).

The use of internal and external genetic markers allows chimeric hosts to be analysed (4, 5). Using them for pole-cell transplantations also makes it possible to distinguish the progeny in subsequent generations (3). For such experiments it is useful to take sterile mutants as hosts (with cell autonomous defects in the germline; for example ovo^{D1}, see reference 33) so that the progeny will only derive from transplanted pole cells.

3.2 Injected markers

The superficial cleavage of most insects allows global labelling of donors by injecting a marker before cellularization. It will be incorporated into all cells when they form later on in development (34). There are several requirements for injected markers to be suitable as cell lineage tracers in transplantation experiments: they should not diffuse into neighbouring cells, they should not have deleterious effects on the cells, they should be passed on to the progeny, and they should reliably stain the whole cell.

3.2.1 Injected enzyme markers

A suitable enzyme marker for injection is horseradish peroxidase (HRP) as it complies with all the requirements mentioned above (for discussion see reference 34; for a detailed description of HRP see reference 35). HRP is a good marker for *Drosophila* embryos as it reliably stains all cells in all tissues (reviewed in reference 1). In postembryonic stages, however, it can be diluted out when cells go on dividing (34) or increase in size (for example larval epidermis). This can be useful if, for example, you intend to differentially label embryonically derived cells in the CNS of postembryonic stages (8).

3.2.2 Injected *in vivo* markers

Throughout embryogenesis it is possible to follow the fate of transplanted cells *in vivo* (8, 34). The only suitable *in vivo* markers known so far are fluorescent dyes. Common markers are fluorescein isothiocyanate dextran (FITC–dextran) and its rhodamine equivalent (RITC–dextran). The molecular weight of the dye should be approximately 20 000 kDa to prevent diffusion through gap junctions. Since cells which are labelled with fluorescent dyes are photosensitized and can be ablated by prolonged excitation (36, 37), you have to lower the irradiation by using special equipment (see Section 7.1). New prospects for labelling techniques will result from the rapid evolution of fluorescent dyes:

(a) There are now fixable fluorescent dyes available (molecular probes; e.g. lysinated fluorescein dextran (38)) which are suitable for cell transplantation (see Section 3.3). Do not use glutaraldehyde for fixation as it emits fluorescent light.

(b) Fluorescence tends to fade within several days without irradiation and within hours during stimulation. Fading can be reduced by adding *p*-phenylene diamine to the mounting medium (39).

(c) Fluorescent label can be transferred into permanent dye by photoconversion (40), or by using specific antibodies against the fluorescent molecule (41).

(d) Minden *et al.* (42) established a method for labelling embryonic nuclei by injecting rhodamine-labelled histones into embryos. This might also be a useful marker for following the number and the fate of transplanted cells.

3.2.3 Experimental design (see also *Figure 3*)

(a) Inject a marker of your choice in the syncytial blastoderm stage (stage 4/5 according to Campos-Ortega and Hartenstein (50)).

(b) Wait until cell closure (stage 6/7 or later) and transplant single cells into unlabelled hosts.

(c) If cells are labelled with fluorescent tracers perform *in vivo* video microscopy during embryogenesis.

(d) Fix the hosts at a stage of your choice and histochemical stain for HRP, or analyse the fixed fluorescent dye, or turn the fluorescent stain into a permanent dye (see above).

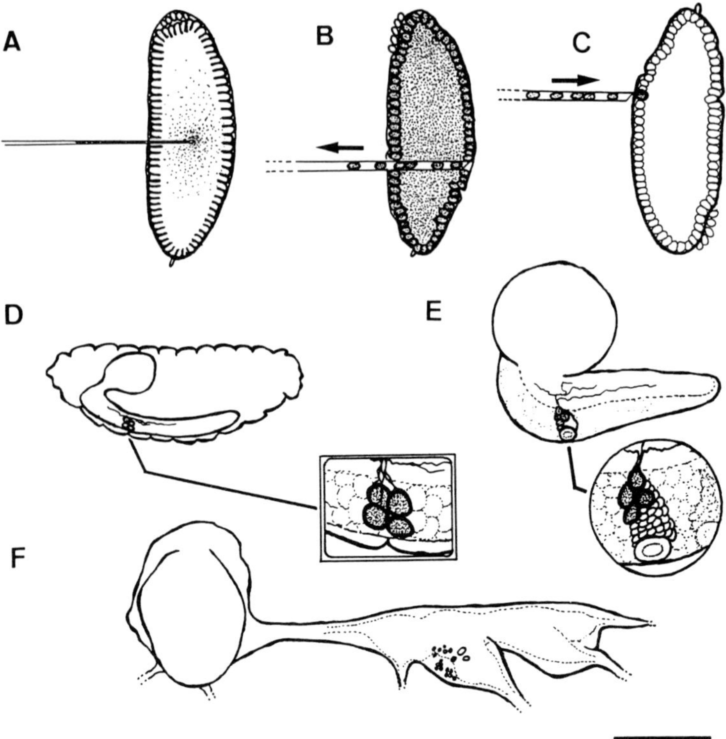

Figure 3. Example of a transplantation experiment using three kinds of markers. (A) A mixture of HRP and RITC–dextran (speckled) is injected into donor embryos during the syncytial blastodermal stage (stage 5) and diffuses throughout the embryo. In addition, the donor strain is gentically labelled with β-gal (thick outline of the cells) at least in all CNS cells. (B) After cell closure (beginning of stage 7) the injected markers become incorporated into all blastodermal cells. Some of these three-fold labelled cells are sucked from ventral thoracic positions into the capillary and are (C) singly transplanted into comparable position in unlabelled host embryos. (D) Clones originating from transplanted cells can be monitored *in vivo* under the fluorescence microscope and documented on a video tape during embryogenesis (see close-up). (E) The hatching larvae are individually reared, the CNS is removed and stained for β-gal and subsequently for HRP (close-up showing a labelled clone). (F) Both markers are still detectable in imaginal stages (taken from reference 8).

3.3 Combination of different markers and staining methods

There are several possible combinations of different markers:

(a) Combinations of internal and external genetic markers (4, 5).

(b) Combinations of fluorescent dyes with HRP (34).

(c) Combinations of genetic markers and injected ones. We were successful in transplanting cells that had been labelled with the following markers at the same time: external genetic markers (*yellow, singed*[3]), internal genetic markers (homozygous insertions of β-gal constructs), injected HRP, and injected RITC–dextran. This combination allowed *in vivo* documentation in the embryo, and histochemical staining for β-gal and HRP at any stage of development (8, *Figure 3*).

(d) Fixable fluorescent dyes allow the simultaneous detection of further stain (e.g. antibody or X-gal staining) in the same cells (the HRP reaction product may be too dark to allow unambiguous codetection).

(e) If you transplant cells carrying homozygous lethal mutations you will face the problem of distinguishing between mutant and wild-type eggs. For this purpose you can make use of blue balancers (43). Choose balancers which carry β-gal constructs with a strong expression in early embryos (for handling see *Protocol 10*). Donors that do not show the expression pattern of the blue balancer are homozygous for the mutation.

4. Embryo preparation

4.1 Egg collection

To guarantee efficient transplantation it is necessary to have sufficient numbers of precisely staged embryos. To achieve this, several strategies have been developed (e.g. references 21, 44). We normally take flies that are older than half a week and younger than two weeks and that have been provided with plenty of fresh yeast. Transfer the flies to plastic cylinders covered at one end with wire mesh (which alows ventilation and prevents condensation inside the cylinder). Place the cylinder with its open end on a Petri dish containing medium (apple juice mixed with red grape juice, 2.5% agar) with a small blob of yeast in the middle. Change the dish every hour. This will provide you with correctly staged embryos within a range of one hour. You can also make use of an automatic egg collector designed by M. Zalokar (19).

4.2 Dechorionation

Protocol 2. Mechanical dechorionation (*Figure 4*; see also
reference 20)

1. Staged eggs are collected from the Petri dish using a small brush or scalpel and transferred to double-sided sticky tape.

2. Fix the tape to a coverslip and place the coverslip on two blocks of agar.

3. Dechorionate the eggs by stabbing with a fine needle between tape and embryo and move carefully along its longitudinal axis until the chorion breaks and releases the embryo.

Protocol 2. *Continued*

4. Pick up the embryos (using the same needle as in step 3) which will individually adhere to the tip of the needle, and transfer up to 100 of them on to one of the blocks of agar medium (deposition on the agar block prevents desiccation of the embryos).

5. Orientate the embryos on the agar block (see Section 4.3).

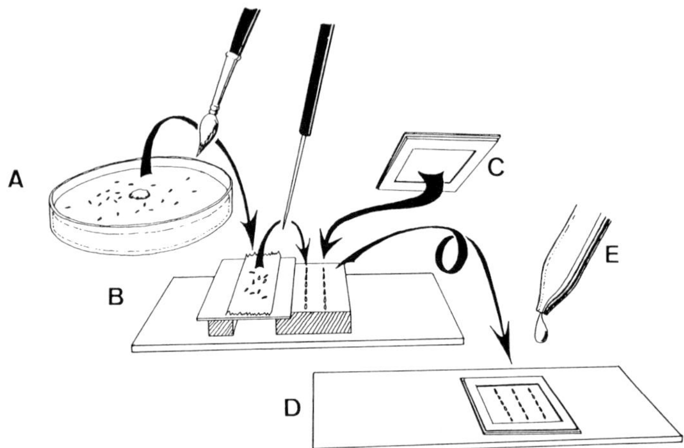

Figure 4. Preparation of embryos for transplantation. Embryos are transferred from the Petri dish to double-sided sticky tape fixed to a coverslip and placed over two blocks of agar (B). The eggs are dechorionated and transferred to one of the agar blocks and orientated. (C) The embryos are covered with a specially prepared coverslip, fixed on a slide with a drop of oil (D), desiccated, then covered with oil (E).

Alternatively, the method described by Santamaria (19) may be used or the chorion may be removed chemically (see also reference 20).

Protocol 3. Chemical dechorionation

1. Gather embryos with a small brush or scalpel and transfer them to a small well containing Ringer's solution (182 mM KCl, 46 mM NaCl, 3 mM $CaCl_2$, 10 mM Tris–HCl pH 7.2; for recipe see reference 31).

2. Exchange Ringer's solution for a 7.5% solution of bleach until the chorion is removed (about 4 minutes, control under a dissecting microscope).

3. Remove the bleach, wash once with Ringer's solution, remove Ringer's as completely as possible with a pulled sharp Pasteur capillary pipette.

4. Transfer dechorionated embryos with a small brush to a block of agar. Embryos can be oriented on the agar block (see Section 4.3).

4.3 Orientation

The position of the transplantation is critical for most of the experiments. It depends on the orientation of the embryos relative to the transplantation capillary (see Section 6.2). Use a fine steel needle to orientate the embryos on the agar block from where they will be stuck directly on to a coverslip (see Section 4.4.1).

4.3.1 Orientation for pole-cell transplantation

Transplantation of pole cells (3, 19) requires an orientation in which the donor embryos are pointing with the anterior, and the host embryos with the posterior pole towards the tip of the transplantation needle (*Figure 5A*). Orientate 50–100 donor embryos parallel to each other with one pole pointing in the same direction. The same is done separately for the hosts. Afterwards stick donors and hosts to separate coverslips. Use a drop of oil to stick the coverslips in opposite directions on a slide (see Section 4.4.1; *Figure 5A*).

4.3.2 Orientation for transplantation of somatic cells

Orientate the embryos with their longitudinal axis perpendicular to the transplantation needle (*Figure 5B*). This makes it possible to penetrate embryos with the capillary at nearly every position along the anterioposterior axis. In addition, orientate the embryos with respect to the ventrodorsal axis, because the transplantation needle is fixed in one position and thus the orientation of

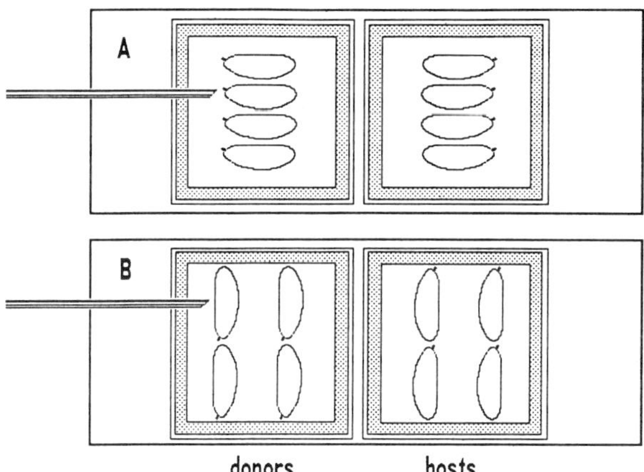

Figure 5. Orientation of donor and host embryos for transplantation. The embryos are fixed to coverslips with a frame of transparent film (stippled). The coverslips are placed on a slide. The transplantation capillary is indicated on the left side. For pole-cell transplantation (A) embryos are orientated parallel to the capillary, donors with the anterior pole (micropyle), hosts with the posterior pole towards the needle's tip. For transplantation of somatic cells (B) the embryos are orientated perpendicular to the needle. In this example cells will be transplanted homotopically in ventral positions (see also *Figures 3* and *6*).

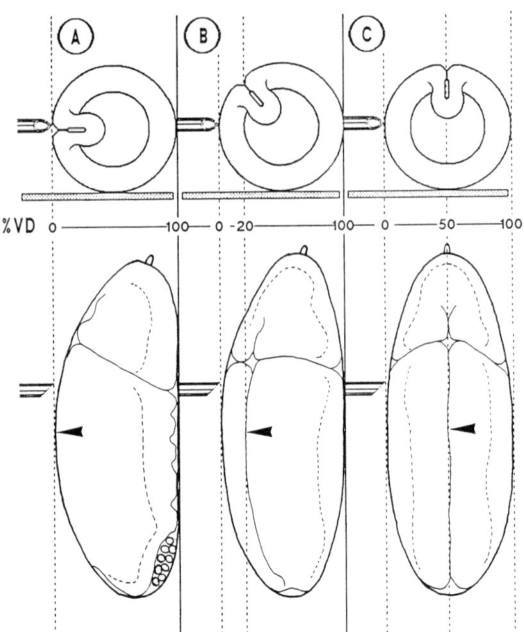

Figure 6. This scheme demonstrates how the orientation of the embryos determines the VD-position of the transplantation. The lower pictures show the microscopic image of embryos, fixed to a slide, orientated with their ventral furrow (arrowheads) at 0%, 20%, and 50% VD (ventrodorsal diameter; see *Figure 7*). The upper pictures show schematic cross-sections of these embryos. The transplantation needle penetrates the embryo on the level of its outer diameter in a position that corresponds to 0%, 20%, and 50% on the two-dimensional fate map (*Figure 7*).

the embryos defines the ventrodorsal position of transplantation (*Figure 6*). The orientation is measured in %*VD* (ventrodorsal diameter), which describes the position of the ventral furrow (= 0% *VD*) in relation to the diameter of the embryo. For selecting the types of precursor cells to be transplanted from/ to the early gastrula stage of *D. melanogaster* a two-dimensional fate map (45) shown in *Figure 7* can be used. The two-dimensional VD-system can be translated to a three-dimensional VDP-system (ventrodorsal perimeter), defining the position on the blastodermal surface that corresponds to several VD-positions (G. Udolph, personal communication; *Figure 7; Table 1*).

Orientate the embryos on the agar block according to the fate map position from/to which you want to transplant cells. Use a fine needle and a dissecting microscope. For embryos in the pregastrula stage there are two criteria that can be used for a rough VD-orientation (see also *Figure 6*):

(a) The embryo is convex on the ventral side and flat dorsally.

(b) Looking from the lateral side, the anterior pole appears tapered with the micropyle sitting slightly dorsally. Looking from the dorsal or the ventral side, the anterior pole appears rounded with the micropyle sitting in the middle.

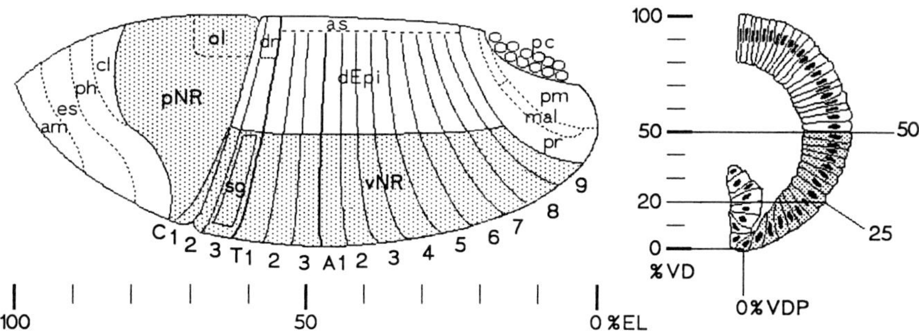

Figure 7. The left scheme shows the two-dimensional fate map of an early stage 7 embryo of *D. melanogaster* (45). In this fate map the position of a cell is defined by its location relative to the longitudinal and dorsoventral axis (given in % egg length (EL) and % of the ventrodorsal diameter (VD), respectively). The right scheme shows one half of a cross-section of a similarly staged embryo at about 50% EL. Using cell injection experiments G.Udolph (personal communication) assigned the two dimensional VD-positions to the position on the three-dimensional blastoderm layer (given in % ventrodorsal perimeter (VDP); see also *Table 1*). A1–9, abdominal segments; am, anterior midgut; as, amnioserosa; C1–3, gnathal segments; cl, clypeolabrum; dEpi, dorsal epidermis; dr, dorsal ridge; es, oesophagus; mal, Malpighian tubes; ol, optic lobe; pc, pole cells; ph, pharynx; pm, posterior midgut; pNR, procephalic neurogenic region; pr, proctodeum; sg, salivary gland; T1–3, thoracic segments; vNR, ventral neurogenic region. The neurogenic region is hatched.

Table 1. Correlation of ventrodorsal diameter (%*VD*), cell number in the vNR (number 1 being the cell lying at the ventral furrow, number 15 being the cell lying at the dorsal margin of the vNR), and ventrodorsal perimeter (%*VDP*) obtained from cell injections by G. Udolph (personal communication).

%*VD*	0	5	20	30	45
Number of injected cells away from the midline	1–2	3–5	6–8	9–10	13–15
%*VDP*	0–6	10–16	20–26	30–36	43–50

After their orientation the embryos will be stuck to a coverslip. Their orientation in the VD-axis will therefore be inverted (see Section 4.4.1). To ease the exact evaluation of the VD-position under the compound microscope later on (see Section 6.2) the ventral side should point towards the lens.

4.4 Fixing embryos to a slide

4.4.1 Sticking the embryos to a coverslip

After the embryos have been orientated on an agar block, they will be fixed to a coverslip (22 × 22 mm). Place a drop of special glue (*Protocol 4*) on it and spread it out with the help of a second coverslip.

Protocol 4. Special glue

1. Fill a 500 ml bottle with as many pieces of brown plastic parcel tape as possible.
2. Fill with about 200 ml *n*-heptane and shake or rotate for several hours to dissolve the glue from the tape.
3. Pour out the heptane solution and centrifuge at about 10 000 r.p.m. (10 000 *g*).
4. The glue supernatant can be concentrated by evaporating or diluted by adding more *n*-heptane.

After the special glue has dried on the coverslip (within a few seconds) fix a frame of single-sided sticky transparent film all around the coverslip (*Figure 4C*) to stop the oil from flowing off the coverslip later on (see Section 4.4.2). Turn the coverslip upside down and press it carefully and vertically on to the agar block with the orientated embryos (*Figure 4*). Do not distort the coverslip when touching the embryos, otherwise their orientation might be altered. Leave a distance of at least 5 mm between the embryos and the edges of the coverslip.

Fix hosts and donors on separate coverslips.

4.4.2 Desiccating the embryos

Desiccate the embryos correctly (*Protocol 5*), so that the egg turgor is low enough to prevent leakage after penetration with the capillary but high enough to guarantee a normal development.

Protocol 5. Desiccation of the embryos

1. Observe the embryos under a dissecting microscope during desiccation. Avoid using a warm light source (an infrared absorbing filter glass can be helpful).
2. When reflection of the light diminishes on the embryos' surfaces you should test the turgor by stroking the embryos with a soft needle (we normally use needles made with steel wire (for springs) having a diameter of 0.3 mm).
3. The turgor is correct when the surface falls transiently into tiny folds while stroking with the needle. The average time for drying is about 10 minutes, but it depends on the temperature, the humidity, and fly strains used. To get more constant conditions Santamaria (19) keeps the drying embryos at a defined temperature in a box containing calcium chloride.
4. When the turgor is correct cover the embryos immediately with a few drops of fluorocarbon oil Voltalef 10S (Atochem) to prevent further drying (46).

4.4.3 Fixing the coverslips to a slide

Use a drop of Voltalef oil to stick the coverslips to a slide. Orientate donors in such a way that the site from which you are going to remove cells will point away from the transplantation needle (you stab through the whole embryo to suck up cells; *Figure 3B*). The hosts, however, must be orientated in the opposite manner with the site of implantation facing the capillary (*Figure 5*).

5. Injection of markers

5.1 Stages for injection

- For transplanting somatic cells inject the marker at the syncytial blastoderm stage or during cell formation (stage 4 or early stage 5; see Section 3.2).

- For pole-cell transplantation (47) inject the marker into the donor embryos before the formation of the pole-cell progenitors (beginning of stage 3).

5.2 The technique (*Protocol 6*)

Protocol 6. Injection of markers

1. Make a solution of 1–10% fluorescent dye and/or HRP in 0.2 M KCl. You can store these solutions as aliquots for several months at −20°C.

2. Place the slide with the embryos under your compound microscope, get them into focus and adjust the tip of the injection needle, with the help of the micromanipulator, to the centre of the microscopic image. When the needle dips into the Voltalef oil the tip will take up a little bit of oil.

3. Pipette about 1 μl of the marker solution on to the Voltalef oil, which covers the donor embryos. Focus your compound microscope on to this drop.

4. Dip the tip of the injection capillary into it and suck up an amount of marker solution that will be sufficient for all donors.

5. Move the needle directly back into the oil (whenever you remove the needle from the oil first suck up a little bit of it to prevent drying of the marker solution at the tip).

6. Adjust the pressure by means of the syringe (see Section 2.3) until the marker solutions reaches into the distalmost tip of the needle, but does not flow out.

7. Get the needle and donor embryos into the same plane. Move the microscope stage to stab the needle centrally into the yolk of a donor embryo (*Figure 3A*). The pressure inside the needle should be high enough to

Protocol 6. *Continued*

inject marker solution but low enough to prevent leakage of the embryo later on. Appropriately desiccated embryos take about 5–10 nl of the solution, depending on their turgor (see Section 4.4.2). For injection, penetrate embryos on the opposite side from which cells will be later taken for transplantation.

8. Make a note about the embryos that have been injected.

---

6. Transplantation

6.1 Stages for cell transplantation

(a) The earliest stage for transplanting somatic cells is about 10 minutes after the onset of germband elongation, when the anterior margin of the posterior midgut anlage has reached dorsally approximately 20% egg length (beginning of stage 7; *Figures 6* and *7*). At 25°C *D. melanogaster* reaches this stage about 170 minutes after fertilization (temperature dependent changes can be taken from reference 21; data for other *Drosophila* species in reference 17). Any later developmental stage may be chosen for donors (13).

(b) Transplantation of pole cells (47) can be carried out from stage 3 (closure of the pole cell progenitors) until stage 7 (when the pole cells move interiorly during amnioproctodeal invagination).

6.2 Positioning the transplantation needle

Transplanting somatic cells requires a very precise positioning of the tip of the transplantation needle.

(a) The VD-position of transplantation is roughly determined by the orientation in which the embryos have been fixed to the slide (see Section 4.3.2). Analyse the position of the ventral furrow under the compound microscope before transplantation (*Figure 6*). Project a per cent scale via a camera lucida into the microscopic image. Alter the magnification of the camera lucida to adapt the scale to the dorsoventral diameter of the respective embryo. Place the needle exactly in the plane in which the outline of the embryo is in focus (see *Figure 6*).

(b) The position on the longitudinal axis is measured in %*EL* (egg length; *Figure 7*). Adjust a per cent scale via a camera lucida on to the longitudinal diameter and place the tip of the needle at the right position.

(c) For pole-cell transplantation orientate the embryos with their poles towards the capillary (*Figure 5*; see Section 4.3.1).

6.3 The transplantation technique (*Protocols 7, 8*)

Protocol 7. Transplantation of somatic cells

1. Fix the transplantation needle to the micromanipulator in such a way that you will see the angle of the tip's bevel during transplantation.

2. Select a donor which is precisely staged (according to Section 6.1). Get the tip into the right position on the blastoderm (according to Section 6.2). Penetrate the donor opposite the side from where cells will be taken. Move the capillary through the embryo until the tip causes a slight bulge in the vitelline membrane on the opposite side (*Figure 3B*).[a]

3. Suck up cells carefully. Take only a few cells, if you want to ensure that you pick up cells in the immediate vicinity of the capillary (at stage 7 one presumptive segment is only about 4 cells wide).

4. Remove the capillary from the donor. Suck up a little bit of oil (to prevent drying of embryonic material at the tip) and lift the needle out of the oil. Get a precisely staged host into the microscopic image and get the needle with the donor cells back into focus.

5. Get one cell[b] near the bevel of the capillary by pressing out yolk and cell fragments distal to it. Clean the outside of the tip by jerking the microscope stage. The yolk will stick to the surrounding oil as small droplets.

6. Move the capillary into the selected blastodermal position of the host and penetrate the embryo so that the bevel of the needle will come to lie slightly beneath the blastodermal cell layer. Move the needle slowly back through the blastoderm while releasing a donor cell out of its tip, so that the cell is placed between the cells of the host.[c] Single cell transplantation can be ensured in three ways:

   - take up only a single cell into the transplantation capillary
   - count the cells in the capillary before and after each transplantation
   - use fluorescent dyes to ensure a single cell transplant in the host a few minutes after transplantation

7. Make a note of which embryos have been used as donors and hosts.

---

[a] Either move the embryos using the microscope stage or move the needle by using the joystick of the micromanipulator.

[b] Uninjured cells inside the capillary show a smooth nucleus in their centre surrounded by granular cytoplasm.

[c] It is important that the implant resides between rather than beneath the blastodermal cells of the host. This will increase the success of integration (for discussion of success rates see reference 34).

Protocol 8. Transplantation of pole cells (see also reference 19)

1. Penetrate donors at the anterior pole and move the capillary through the yolk until you reach the pole cells.
2. Suck them up and move to the hosts. Pole cells are distinguishable from somatic cells in the capillary, as they are less granular.
3. Penetrate the hosts from the posterior pole and release a number of cells (which will depend on the type of experiment) between the pole cells of the host.

---

7. Processing the hosts

7.1 *In vivo* documentation

To avoid ablation of fluorescently labelled cells (see Section 3.2.2), *in vivo* documentation has to be performed with low-intensity dye irradiation coupled with a high-sensitivity video camera. In combination with a suitable time-lapse recorder and a shutter triggered by a time-controller it is possible to document the behaviour of the labelled cells while considerably reducing the total time of their exposure to fluorescent light. The documentation may be further improved by image-processing systems (for further details see reference 48).

7.2 Rearing hosts (*Protocol 9*)

Protocol 9. Rearing hosts

1. Remove or destroy those embryos on the coverslip that have not been injected, with the help of a sharp needle under a dissecting microscope.
2. Transfer the coverslip with the embryos into a disposable weighing dish (size of the flat bottom 28 × 28 mm).[a]
3. Fill with Voltalef oil (3S) so that the embryos at the bottom of the dish are covered by about 2 mm oil (to prevent desiccation).
4. Put the weighing dish into a moist chamber.
5. Allow embryos to further develop at a suitable temperature (16–25°C depending on the temporal experimental requirements) until they reach the stage of your choice.
6. Pick hatched larvae from the weighing dish with the help of a sharp needle and transfer them to standard medium.

---

[a] Alternatively, you may leave the embryos on the slide and put them into a moist chamber (20), or put the coverslip with the embryos directly on to a Petri dish with standard medium and let them hatch (19).

7.3 Preparation procedures (*Protocols 10–12*)

At stage 16, embryos have almost completed their differentiation but the cuticle is still thin and permeable. This allows the embryos to be fixed easily.

Protocol 10. Preparation of embryos not older than stage 16 (modified from reference 49)

1. Start the preparation when the embryos have reached the appropriate stage. Embryos older than stage 16 have to be treated according to *Protocol 11*. Stage 16 embryos can be recognized under a dissecting microscope since their midgut will have become constricted and shows up three portions.

2. Prepare fixative from one part *n*-heptane and one part 25% glutaraldehyde[a] and shake it vigorously.[b] For fixation use only the heptane phase.

3. Pour off the Voltalef oil from the weighing dish (*Protocol 9*) (can be reused many times).

4. Cover the embryos with fixative (heptane phase) and wait until embryos start turning yellow (about 10 minutes at room temperature).

5. Pour off the fixative and wash several times with 0.2 M phosphate buffer (PB) or phosphate-buffered saline (PBS; see reference 31 for recipe).[c] Add about 100 µl of any serum to the buffer during the last washing step, so that the embryos will not stick to the coverslip or the capillary after devitellinization. (PBS: 130 mM NaCl, 7 mM Na_2HPO_4, 3 mM NaH_2PO_4 pH 7.2.)

6. Take a pulled capillary, break off the tip and use it to scratch the vitelline membrane and release the embryos. Devitellinizing the embryos on the coverslip allows individual treatment of each embryo according to your experimental needs.

7. Transfer the devitellinized embryos with a pipette to a small well containing buffer, proceed with further staining.

[a] If you use formaldehyde instead of glutaraldehyde the mechanical devitellinization may be more difficult. In this case you may alternatively devitellinize the embryos chemically using methanol (31).
[b] It can either be stored in the refrigerator up to two weeks or prepared freshly.
[c] At this point of the protocol you can stain for blue balancers (see Section 3.3). Cover the embryos with X-gal staining solution (*Protocol 14*) and scratch the vitelline membrane a little bit to accelerate the penetration with X-gal.

Stage 17 embryos are fully differentiated and thus have completed cuticle secretion. Therefore, the fixative has to be injected.

Protocol 11. Preparation of stage 17 embryos

1. Use a very sharp injection needle to inject a 4% solution of paraformaldehyde in PEM buffer (0.1 M Pipes, 2 mM $MgSO_4$, 1 mM EGTA, pH 6.9) under the compound microscope (for injection procedure see *Protocol 6*).
2. Wait for 10 minutes, remove the oil, wash with PBT, and finally with buffer. Devitellinize the embryos and transfer them with a pipette to a depression slide containing buffer.
3. Cut the tips off the embryos, in order to allow penetration of staining solutions. Seize the distalmost tips with sharp forceps and cut using a splinter of a razor blade along the shanks of the forceps.
4. Transfer the embryos with a Pasteur pipette to a small well or microfuge tube to stain further.

There are many methods of dissecting larval and imaginal tissues (21). The following method allows preparations of all tissues of late larvae to be made without distorting their typical shape and location.

Protocol 12. Preparation of larvae

1. Immobilize larvae at 4°C and dry them on filter paper.
2. Fix them with special glue (*Protocol 4*) to a cooled slide.
3. Inject a solution of 1% glutaraldehyde or 4% paraformaldehyde in any buffer under a dissection microscope. Use a pulled and broken capillary. Inject applying high pressure, so that the larva expands.
4. Wait about one minute, retract the capillary, transfer the larva to a drop of Ringer's solution (31) and pin it at its distalmost tips to a wax dish. Use fine tungsten dissecting needles.
5. Take a fresh splinter of a razor blade and cut along the longitudinal axis of the epidermis.
6. Pin down the epidermis at the sides and remove all tissues that you are not interested in.
7. Postfix with the same fixative as in step 3, wash in buffer, and continue with the chosen staining procedure.

8. Staining procedures

There are staining protocols available for a variety of enzymatic markers (for example, references 21, 31). We will focus on two widely-used enzymatic markers, HRP and β-gal:

Protocol 13. Staining for HRP

1. Fix embryos according to *Protocols 10* or *11*. Fix dissected tissues for about 15 minutes with 1–2.5% glutaraldehyde in PB or PBS. Wash subsequently with buffer.

2. Dissolve 0.5–1 mg of diaminobenzidine (DAB)[a,b] per 1 ml of PB or PBS. Freeze and store it in aliquots for up to several months.

3. Add 1 μl of 30% H_2O_2 to 1 ml DAB solution.

4. Remove PB or PBS from the fixed specimen and add the DAB/H_2O_2-staining solution. Wait for about 5–10 minutes. Positive staining can be recognized under the dissecting microscope.

5. Remove staining solution.

6. Either wash once with PB or add 70% alcohol directly for dehydration.

[a] For alternatives (e.g. chloronaphthol) see reference 35.
[b] DAB is carcinogenic and has to be disposed of separately from normal refuse and sewage, follow all recommended precautions.

Protocol 14. Staining for β-gal

1. Prepare the oxidation buffer (150 mM NaCl, 1 mM $MgCl_2$, 3.3 mM $K_3[Fe(CN_6)]$ in 0.2 M PB, pH 7.2) and the X-gal stock solution (20% X-gal in dimethyl sulfoxide). X-gal stock solution can be stored in aliquots in the freezer for up to several months.

2. Fix embryos according to *Protocols 10* or *11*. Fix dissected tissues for about 10 minutes with 1% glutaraldehyde in PB or PBS. Wash subsequently in oxidation buffer.

3. Warm up 1 ml of oxidation buffer to about 60°C, add 10 μl of X-gal stock solution, shake vigorously and let the staining solution slowly cool down to room temperature (otherwise X-gal will precipitate).

4. Remove buffer from the fixed specimen and add staining solution.

5. Incubate for several hours at 37°C.

6. Remove staining solution and start dehydration, or, if necessary, rinse with buffer and stain for a second marker (for example HRP).

Antibody staining for β-gal is also a common method (31). However, antibodies require fixing in formaldehyde or paraformaldehyde. This kind of fixation is less easy to deal with after transplantation (see *Protocol 10*). Therefore, only carry out antibody staining when it is absolutely necessary (for example double staining with fixable fluorescent dyes; see Section 3.3).

9. Embedding the preparations

Araldite or Epon are suitable, at least for embedding preparations stained with DAB, X-gal and *p*-nitro blue tetrazolium-chloride (NBT).

Protocol 15. Transfer to embedding medium (suitable for whole mounts)

1. Dehydrate tissues by transferring them to 70% ethanol (1 minute is sufficient) and subsequently to dry 100% ethanol for 10 min. Change the 100% ethanol once or twice. Ensure that the edges of the dish are free of water drops.
2. Remove alcohol and add a few drops of xylene.
3. Wait up to a minute until the tissue is cleared, remove most of the xylene, and add the embedding medium from a syringe.[a]
4. Leave for about one hour under a hood until the rest of the xylene has evaporated.

[a] The embedding medium can be stored in single-use syringes at $-20\,°C$ for years.

Protocol 16. Embedding on slides (sandwiches)

1. Take a fine needle and transfer the preparations singly in a little drop of embedding medium to a large coverslip (22×60 mm).
2. Orientate them in the drops, so that the side of interest will face the lens.
3. Polymerize at $70\,°C$ and check the orientation of embryos every 15 min until the plastic has hardened.
4. Cover with embedding medium and a small coverslip and polymerize at $70\,°C$.
5. Fix the coverslip-sandwich with a tape either side up on a slide.

Preparations can be embedded in capillaries and visualized from all sides under the compound microscope. Use 65 mm long borosilicate glass capillaries (Hilgenberg) with an internal diameter of 0.2 mm for embryos, CNS of early larvae, and ventral nerve cords of late larvae, and capillaries of an internal diameter of 0.24 mm or 0.32 mm for imaginal CNS or late larval hemispheres.

Protocol 17. Embedding in capillaries

1. Suck preparations into the capillary. Sometimes it is necessary to push them with a hair fixed to a small holder.

2. Seal the capillary by sticking the tip carrying the preparations into plasticine.

3. Clean the outside of the capillary thoroughly with acetone.

4. Fix a very small flag made of tape to the other end of the capillary, this will be used to rotate the capillary later on (*Figure 8*).

5. Fix larger capillaries (see *Protocol 1*) with a drop of Araldite or Epon to a slide. Store your preparation-capillaries inside these larger capillaries (*Figure 8*).

6. Pull a polyethylene hose (see Section 2.3), cut a 1 cm section from the thinnest part and fix it on to one end of a slide.[a] For inspection under the microscope insert the preparation-capillary into the short hose (*Figure 8*). For microscopy use a drop of immersion oil even for low magnification. Alter the position of the specimen by rotating the capillary with the flag.

[a] Polyethylene will prevent damage to the lens if you scratch it accidentally.

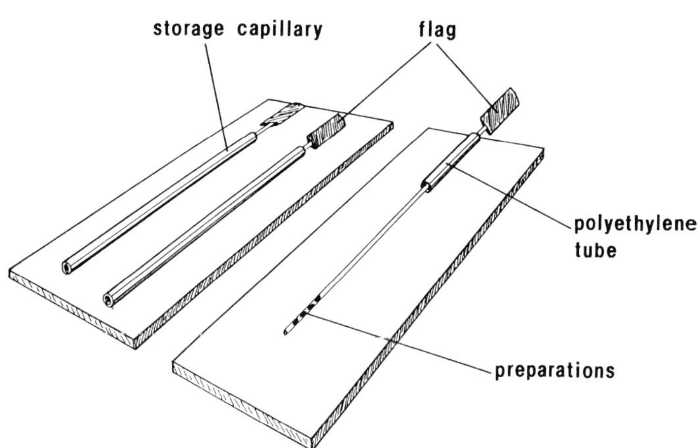

Figure 8. Capillary holders. For storage, preparation-capillaries are inserted into larger capillaries (about 60 mm in length) which have been fixed to glass slides (left). For microscopy, a piece of polyethylene tubing is fixed to a slide to take up the preparation-capillary (right). A piece of tape (flag) attached to the end of the preparation-capillary allows it to be turned easily.

Acknowledgements

Firstly we thank all the former and present members of our group for permanent innovations in the technique. We thank especially Thomas Becker, Torsten Bossing, and Karin Lüer. We thank Gerald Udolph for putting his fate-mapping data at our disposal. We thank Hazel Smith and Claudia Sachs for critically reading the manuscript.

References

1. Technau, G. M. (1987). *Development,* **100,** 1.
2. Gardner, R. L. (1968). *Nature,* **220,** 596.
3. Illmensee, K. (1973). *Roux's Arch. Dev. Biol.,* **171,** 331.
4. Illmensee, K. (1976). *Experimental Cell Research,* **97,** 127.
5. Illmensee, K. (1978). In *Genetic mosaics and cell differentiation* (ed. W. J. Gehring), p. 51. Springer-Verlag, Berlin.
6. McConnell, S. K. (1985). *Science,* **229,** 1268.
7. Blakemore, W. F. and Franklin, R. J. M. (1991). *Trends Neurosci.,* **14,** 323.
8. Prokop, A. and Technau, G. M. (1991). *Development,* **111,** 79.
9. Snyder, W. Y., Deitcher, D. L., Walsh, C., Arnold-Aldea, S., Hartwieg, E. A., and Cepko, C. L. (1991). *Cell,* **68,** 33.
10. Simcox, A. A. and Sang, J. H. (1983). *Dev. Biol.,* **97,** 212.
11. Technau, G. M. and Campos-Ortega, J. A. (1986). *Roux's Arch. Dev. Biol.,* **195,** 445.
12. Jacobson, M. and Xu, W. (1989). *Dev. Biol.,* **131,** 119.
13. Technau, G. M., Becker, T., and Campos-Ortega, J. A. (1988). *Roux's Arch. Dev. Biol.,* **197,** 413.
14. Müller, C. M. and Best, J. (1989). *Nature,* **342,** 427.
15. Technau, G. M. and Campos-Ortega, J. A. (1987). *Proc. Natl Acad. Sci. USA,* **84,** 4500.
16. Rosenberg, M. B., Friedmann, T., Robertson, R. C., Tuszynski, M., Wolff, J. A., Breakefield, X. O., and Gage, F. H. (1988). *Science,* **242,** 1575.
17. Becker, T. and Technau, G. M. (1990). *Development,* **109,** 821.
18. Le Douarin, N. M., Renaud, D., Teillet, M. A., and Le Douarin, G. H. (1975). *Proc. Natl Acad. Sci. USA,* **72,** 728.
19. Santamaria, P. (1986). In Drosophila: *a practical approach* (ed. D. B. Roberts), p. 159. IRL Press, Oxford.
20. Spradling, A. C. (1986). In Drosophila: *a practical approach* (ed. D. B. Roberts), p. 175. IRL Press, Oxford.
21. Ashburner, M. (1989). Drosophila: *a laboratory handbook.* CSH Laboratory Press.
22. Zalokar, M. (1981). *Experientia,* **37,** 1354.
23. Lawrence, P. A., Johnston, P., and Morata, G. (1986). In Drosophila: *a practical approach* (ed. D. B. Roberts), p. 229. IRL Press, Oxford.
24. Parker-Thornburg, J. and Bonner, J. J. (1987). *Cell,* **51,** 763.
25. Hiromi, Y., Kuroiwa, A., and Gehring, W. J. (1985). *Cell,* **43,** 603.
26. Simon, J. A., Sutton, S. C., Lobell, R. B., Glaser, R. L., and Lis, J. T. (1985). *Cell,* **40,** 805.
27. O'Kane, C. and Gehring, W. J. (1987). *Proc. Natl Acad. Sci. USA,* **84,** 9123.
28. Bellen, H. J., O'Kane, C. J., Wilson, C., Grossniklaus, U., Pearson, R. K., and Gehring, W. J. (1989). *Genes Dev.,* **3,** 1288.
29. Bier, E., Vässin, H., Shepherd, S., Lee, K., McCall, K., Barbel, S., Ackerman, L., Carretto, R., Uemura, T., Grell, E., Jan, L. Y., and Jan, Y. N. (1989). *Genes Dev.,* **3,** 1273.
30. Wilson, C., Pearson, R. K., Bellen, H. J., O'Kane, C., Grossniklaus, U., and Gehring, W. J. (1989). *Genes Dev.,* **3,** 1301.

31. Ashburner, M. (1989). Drosophila: *a laboratory manual*. CSH Laboratory Press.
32. Knipple, D. C. and MacIntyre, R. J. (1984). *Mol. Gen. Genet.*, **198**, 75.
33. Lindsley, D. L. and Zimm, G. (1990). Drosophila *Information Service, **68**, 1.
34. Technau, G. M. (1986). *Roux's Arch. Dev. Biol.*, **195**, 445.
35. Nässel, D. R. (1983). In *Functional neuroanatomy* (ed. N. J. Strausfeld), Springer-Verlag, New York.
36. Miller, J. P. and Selverston, A. I. (1979). *Science*, **206**, 702.
37. Shankland, M. and Weisblat, D. A. (1984). *Dev. Biol.*, **106**, 326.
38. Gimlich, R. L. and Braun, J. (1985). *Dev. Biol.*, **109**, 509.
39. Johnson, G. D., Gloria, M., and Araujo, C. N. (1981). *J. Immunobiological Methods*, **43**, 349.
40. von Bartheld, C. S., Cunningham, D. E., and Rubel, E. W. (1990). *J. Histochem. Cytochem.*, **38**, 725.
41. Taghert, P. H., Bastiani, M. J., Ho, R. K., and Goodman, C. S. (1982). *Dev. Biol.*, **94**, 391.
42. Minden, J. S., Agard, D. A., Sedat, J. W., and Alberts, B. M. (1989). *J. Cell Biol.*, **109**, 505.
43. Klämbt, C., Jacobs, J. R., and Goodman, C. S. (1991). *Cell*, **64**, 801.
44. Roberts, D. B. (ed.) (1986). Drosophila: *a practical approach*. IRL Press, Oxford.
45. Technau, G. M. and Campos-Ortega, J. A. (1985). *Roux's Arch. Dev. Biol.*, **194**, 196.
46. Zalokar, M. (1981). *Microscopica Acta*, **84**, 231.
47. Technau, G. M. and Campos-Ortega, J. A. (1986). *Roux's Arch. Dev. Biol.*, **195**, 489.
48. Weiss, D. G., Maile, W., and Wick, R. A. (1989). In *Light microscopy in biology: a practical approach* (ed. A. J. Lacey), p. 221. IRL Press, Oxford.
49. Zalokar, M. and Erk, I. (1977). *Stain Technol.*, **52**, 89.
50. Campos-Ortega, J. A. and Hartenstein, V. (1985). *The embryonic development of Drosophila melanogaster*. Springer-Verlag, Berlin.

3

Clonal analysis in plants

MARK DUDLEY and R. SCOTT POETHIG

1. Introduction

Little is known about the nature of cell–cell interactions in plants as many of the most useful methods for studying this problem are difficult, if not impossible, to employ in plants. One of the most significant technical problems is the tendency of plant cells and tissues to dedifferentiate in new environments. As a result, methods that are routinely used to examine cellular interactions in many other systems, such as tissue grafts, cell mixing experiments, or microsurgical manipulations, are not generally useful for studying cellular interactions in plants (although they have been used to study interactions between tissues or organs). In order to study the cellular interactions that regulate cellular and multicellular patterns of cellular differentiation, these interactions must be studied *in situ*, with a minimum of experimental manipulation. The ability to juxtapose cells with different developmental potentials in genetic mosaics is one of the few methods by which this can be achieved in plants.

Genetic mosaics have been used to address two types of questions about the mechanism of plant development. The first is that of cell fate. Mutations with visible, cellular phenotypes have been used as cell lineage markers in a number of different species to define the fate of cells in developing primordia (1–4). This approach does not provide specific information about the mechanism regulating cell fate, but does reveal some of the cellular parameters of morphogenetic processes, e.g. the number and fate of the founder cells in a primordium, and the rate and orientation of cell division. Genetic mosaics have also been used to investigate gene function by linking marker genes with genes that have morphological phenotypes. In this case, the question is whether the morphological gene is expressed cell-autonomously or noncell-autonomously. A gene is cell-autonomous if its expression is required in the cells that manifest the trait regulated by that gene. Nonautonomous expression implies that cells and tissues which display a particular phenotypic trait need not be the same cells which express the gene for that trait. Cell-autonomous genes are predicted to encode gene products that are confined to individual cells, whereas nonautonomous genes are likely to encode diffusible factors, or extracellular signals involved in cell–cell interactions. In this chapter we will focus specifically on the use of genetic mosaics for this latter type of analysis. For more extensive discussions of the origin and use of genetic mosaics in plants see references 1–7.

1.1 Types of genetic mosaics

All the tissue in the shoot of a flowering plant ultimately arises from cells of the shoot apical meristem (SAM). In flowering plants, the pattern of cell division within the SAM produces three distinctive cell lineages. The two outer cell layers of the SAM (L1 and L2) remain as single cell layers because they divide exclusively in an anticlinal plane; the third cell layer (L3) contributes to the bulk of the tissue in the shoot by dividing in both anticlinal and periclinal planes. Each of these lineages arises from approximately three cells located at the tip of the SAM.

Genetic mosaics are usually classified according to the genetic constitution of the initial cells of SAM. Plants in which the initial cells of the L1, L2, and L3 are genetically homogeneous, but where one lineage is genetically different from the others, are called *periclinal chimeras*. Periclinal chimeral shoots are genetically stable because cells from one layer rarely invade another cell layer. Moreover, because the stratification of the primary SAM is retained during the initiation of lateral buds, periclinal chimeras can be perpetuated more-or-less indefinitely from lateral branches. Mosaics in which a cell layer is genetically heterogeneous eventually lose their mosaic structure because the composition of the initial cell population at the tip of the SAM changes during shoot growth. These mosaics are of two types: *sectorial chimeras*, in which a sector encompasses a fraction of the cells in all three layers; and *mericlinal chimeras*, in which a sector is confined to a single cell layer. Of course, most genetic mosaics do not originate from events involving the initial cells of the shoot meristem because these cells represent a very small fraction of the cells in the shoot. Spontaneous genetic mosaics, and those induced by various mutagenic treatments, commonly originate from changes in the genotype of cells located at the periphery of the SAM, or in lateral organs. This type of mosaicism is short-lived because peripheral cells are eventually displaced from the meristem by the products of the apical initials. However, peripheral sectors (or 'extra-apical' lineages as they are sometimes called) may produce stable, chimeral shoots if a lateral shoot happens to form within such a sector.

1.2 Creating genetic mosaics

When genetic mosaics are used to investigate the autonomy of gene expression, the goal is to obtain sectors that differ genetically from other tissue in the plant with respect to:

(a) a gene with an interesting developmental or physiological phenotype, whose *a*utonomy is *u*ndergoing *t*esting (the AUT gene);

(b) a visible *m*arker gene (M) that allows identification of the sector and delineates the sector's boundaries.

Genetic mosaics possessing such sectors can be produced by:

(a) inducing mutations or chromosomal rearrangements in somatic cells of an intact plant;

(b) regenerating shoots from genetically heterogeneous cell mixtures;

(c) mutations that block nuclear fusion during fertilization.

No single approach is generally applicable. The first approach requires that the M and AUT genes be located on the same chromosome. Although it is theoretically possible to accomplish this in any species, this requirement actually limits this approach to species with reasonably good genetics. The cell-mixing approach has less stringent genetic requirements in that the M and AUT genes do not have to be genetically linked; however, this approach can only be used with species that regenerate well in tissue culture or *in situ* (i.e. produce adventitious buds), and for this reason, success has been limited primarily to species in the family Solonaceae. The last approach also does not require linkage between the M and AUT genes, but depends on the existence of a mutation that blocks syngamy. Since cotton is the only species in which such a mutation has been described (8), this method will not be discussed further.

2. Graft chimeras

The first genetic mosaics to be produced intentionally in plants were interspecific chimeras produced by grafting shoots of tomato (*Lycopersicon esculentum*) and black nightshade (*Solanum nigrum*) and encouraging adventitious shoot production in the region of the graft joint (see references 6, 9). This technique has been used more recently to produce interspecific chimeras in the genus *Nicotiana* (10), and inter- and intraspecific chimeras involving tomato and *Solanum* species (11, 12). It is also possible to obtain interspecific chimeras *in vitro* from mixed cell cultures (13–15). However, a comparison of the frequency of chimeras obtained in tissue culture and by a protocol similar to the one described below (*Protocol 1*) suggests that the latter method is more effective (10).

Chimeral shoots have been historically recognized as shoots that have the epidermal phenotype of one parent and the morphology of the other. More recently, the identification of chimeral shoots has been facilitated by using parents that differ with respect to an obvious epidermal trait and a visible trait expressed in internal tissue. Mutations that affect the density of epidermal hairs and the pigmentation of internal tissue (e.g. a yellow–green mutation) have been particularly useful. In fact, unless one of the parents of a chimera is chlorophyll-deficient, it is extremely difficult to identify chimeras in which the

L3 differs genetically from the L1 and L2. It is convenient to develop a single marker stock that carries both of these mutations so that M genes do not have to be introduced into other species or stocks carrying AUT mutations.

Protocol 1. Producing graft-chimeras

1. Reciprocally graft shoots when they are approximately 30–50 cm tall. Make a diagonal cut through the middle of the stem of both parents, leaving several leaves below the cut. Remove all but the smallest leaves on the scion and attach it to the root stock by wrapping the graft joint several times with a strip of Parafilm or with budding rubbers.

2. Keep the grafted shoots in a humid environment (e.g. under a plastic bag or in a mist chamber) for 10–21 days, or until the plants are firmly grafted.

3. After the graft region is well-joined, cut the scion off above the graft joint, leaving a 1–2 mm section of the stem on the root stock. Coat the cut surface with lanolin to prevent drying.

4. Keep the plants in a greenhouse or growth chamber, and harvest shoots from the graft joint as they appear. Also remove lateral branches that arise from axillary buds because these tend to inhibit adventitious bud formation at the graft joint. Examine each new adventitious shoot carefully for the presence of a sector, particularly near the base of the shoot.

5. Root mosaic shoots, and prune them to encourage the outgrowth of lateral buds in the region of the sector. By carefully selecting buds that arise at the boundary of a sector it is often possible to obtain periclinal chimeras of several different types.

---

3. Clonal analysis

In species with well-developed genetics, one of the most useful ways to investigate the autonomy of gene expression is by clonal analysis. In contrast to the method described above, clonal analysis involves the use of techniques that produce changes in the genotype of somatic cells during normal shoot growth. Genetic diversity at the AUT gene is generated in a single cell, for example in the shoot apical meristem, by agents that induce chromosomal rearrangements or by spontaneous chromosome loss. The clonal descendants of this cell form a sector of tissue in the mature plant that is genetically different at the AUT locus from surrounding tissues. Usually, the sector is identified in the mature plant by the coordinate expression of a linked M gene with an obvious cell-autonomous phenotype. When a sector expressing the M gene is found, the phenotype of the AUT gene is assayed in the sector. If the sector expresses the phenotype characteristic of the AUT genotype of the sector (i.e. its phenotype is different from surrounding tissues), then the AUT gene is cell-autonomous. If the sector expresses the phenotype characteristic of surrounding tissues, or if the phenotype of surrounding tissues is modified by the sector, the gene is noncell-autonomous.

3.1 Genetic requirements for assaying cell autonomy by clonal analysis

The genetic requirements for the generation of sectors appropriate for the analysis of cell autonomy are illustrated in *Figure 1*. There are four, more-or-less generic, requirements:

(a) The tissue must be heterozygous for a mutation of the AUT gene, or it must have the potential for establishing sectors with genetic diversity at the gene of interest (e.g. by the excision of a transposable element).

(b) The recessive allele of the AUT gene needs to be linked in *cis* to the recessive allele of a scorable M gene.

(c) The M gene and the AUT gene must be positioned on the chromosome, relative to one another and to the centromere, in such a way that re-arrangements which uncover the M gene also uncover the recessive allele of the AUT gene.

(d) Because sectors may be quite small, the phenotype conferred by the AUT gene must be visible in small groups of cells or in sectors that occupy a fraction of a mature organ.

Different types of M genes are compared in Section 3 and methods of generating sectors are described in detail in Section 3.2. The order of genes shown in *Figure 1*, i.e. centromere—M—AUT, can be critical to this approach because chromosome breaks or recombination events will only uncover genes that are distal to the site of these events. When the location of these events is not predictable (e.g. in gamma-ray induced sectors), only

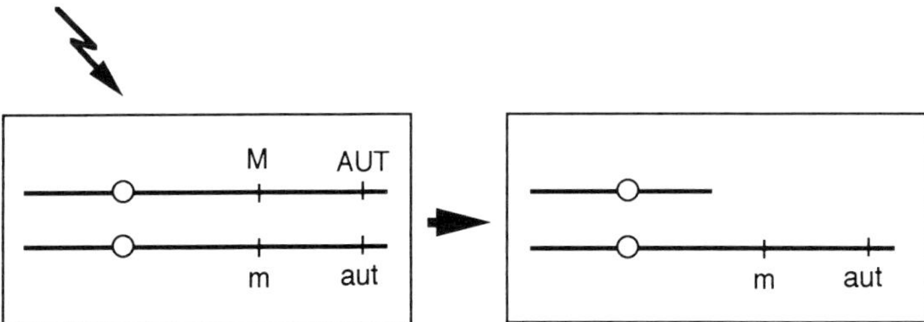

Figure 1. Genetic requirements for marking cell lineages by techniques that induce chromosome loss. The recessive allele of the M gene (*m*) must produce a visible, cell-autonomous phenotype and must be linked to the recessive allele of the AUT gene (*aut*). When ionizing radiation is used to induce chromosome breakage, it is desirable for the M gene to be positioned proximally to the AUT gene so that breaks between M and the centromere necessarily uncover both *m* and *aut*.

proximal–distal linkage of the M gene and the AUT gene will result predominantly in marked sectors that also express the recessive allele of the AUT gene. If the AUT gene is located proximal to the M gene, breaks between the two loci will uncover the M gene without uncovering the AUT gene and the resulting sector will display the dominant AUT phenotype, regardless of the autonomy or non-autonomy of this gene.

Although it is often possible to identify an appropriate M gene located proximal to a gene of interest, in some cases no such M gene is available. One way to circumvent this problem is to link the AUT gene to an M gene on another chromosome using a reciprocal translocation. Maize is particularly amenable to this approach because a large number of reciprocal translocations exist which link most of the genome to useful marker mutations. Alternatively, if an M gene is located in a distal position relative to the AUT gene, then it may be possible to reverse the order of these genes using a paracentric inversion. If the inversion includes the dominant alleles of the M and AUT genes, then breaks that occur proximal to the M gene on the inversion chromosome will result in the loss of the dominant AUT allele as well.

3.2 Generating stocks for clonal analysis

No single series of genetic crosses will be optimal for every potential AUT gene. The particular crosses required to generate useful stocks will depend on the vagaries of the particular gene under investigation. Factors that must be considered include:

- whether the AUT mutation is dominant or recessive
- whether the M gene is dominant or recessive
- whether the M or AUT genes have a homozygous lethal or sterile phenotype
- the genetic distance between the M and AUT loci
- the difficulty of assessing the AUT phenotype

Clearly, there is a wide array of genetic strategies that could prove useful for various combinations of AUT genes and M genes. In addition to the fairly obvious genetic components listed above, experience has taught us that a variety of other, sometimes subtler issues can and must be addressed from the very first cross.

(a) Obtaining enough seeds to generate a large number of sectors is important and may be as difficult as obtaining stocks of the appropriate genotype.

(b) If the M–AUT stock must be propagated in heterozygous condition (because either the M or AUT mutation is lethal or sterile), it may be necessary to test the genotype of plants with somatic sectors to ensure that these genes have not recombined.

(c) When possible, multiple methods for generating or marking mosaic plants should be pursued. This is often helpful because 'background' genetic effects can obscure the phenotype of the AUT gene, and because certain types of M genes are visible in one tissue and not another.

(d) Generation of stocks to map the M and AUT genes should be pursued simultaneously with the generation of stocks to produce sectors. Where these genes are closely linked, but their relative order has not been unambiguously determined, mapping is particularly important. The 'wrong' orientation can generate marked sectors that are interpreted as nonautonomous, but which are heterozygous at the AUT locus and are, therefore, not relevant for determining the autonomy of this gene. When a reciprocal translocation is used to link an AUT gene to a proximal M gene, it is particularly critical to map these genes relative to the translocation breakpoint. This is necessary not merely to ensure that these genes are in the proper orientation, but to guard against errors in the identification of the chromosomes involved in the translocation.

3.3 Marker genes

The choice of the optimal marker mutation depends on the particular gene undergoing analysis. In general, an 'ideal' M gene should be:

- well mapped relative to the AUT gene (usually proximally linked)
- cell autonomous, so the boundaries of the sector are clearly distinguishable
- easy to score, so that rare sectors can be identified among a large number of unsectored plants
- expressed in the tissues and organs which are affected by mutations at the AUT locus
- irrelevant to the expression of the AUT phenotype
- as far as possible from the centromere, because the probability of breaks between the M gene and the centromere (in cases where ionizing radiation is used to generate sectors) is directly proportional to this distance

Several major classes of M genes have been used for clonal analysis in plants. Photosynthetic (e.g. albino) mutations and anthocyanin mutations are described below (Sections 3.3.1 and 3.3.2, respectively). Other mutations that have been used as markers are mentioned in the 'miscellaneous' markers section (Section 3.3.3).

3.3.1 Photosynthetic mutations

Photosynthetic mutations represent the most fully exploited and often the most useful class of markers, for the following reasons:

(a) They are generally cell autonomous and have a readily visible (white or pale green) phenotype.

(b) Numerous loci affect photosynthetic reactions—everything from reaction centre enzymes to genes for chloroplast assembly and development.

(c) These loci are scattered throughout the genome, so it is likely that a useful photosynthetic M will be linked to any given AUT gene

(d) Most are hemizygous viable (in a sector), and some are homozygous viable.

(e) They are expressed in all photosynthetically active tissues, including leaves, stems, and bracts.

(f) They are expressed throughout development in basal (juvenile) structures which senesce during plant growth, and in later (adult) structures present at plant maturity.

(g) Derivatives of all meristem layers—L1, L2, and L3—can produce photosynthetically active tissue, and thus, identifiable sectors.

Although photosynthetic mutations are the most widely used type of M gene, they have certain disadvantages. Because only a few cells in the epidermis produce chlorophyll, sectors that are confined to this tissue are difficult to identify. Epidermal sectors can usually only be detected by examining the autofluorescence of guard cell chloroplasts in epidermal peels. Photosynthetic mutations are also of little value for marking cell lineages in achlorotic tissues such as roots and petals. It should also be noted that albino mutations are homozygous lethal, requiring that all crosses be made with stocks that are heterozygous for these mutations. Unless recombination between an albino M gene and the AUT gene can be suppressed or minimized, recombination in heterozygous plants during the amplification of stocks for clonal analysis may make the AUT genotype of any particular sector ambiguous.

3.3.2 Anthocyanin mutations

Anthocyanin mutations have been used for clonal analysis most effectively in maize, where the genetic regulation of anthocyanin is well understood (16). However, many plants produce anthocyanins in various plant parts, and mutations affecting anthocyanin production are not uncommon. Some of the advantages of anthocyanin mutations are shown below:

(a) Multiple genes are required for anthocyanin production. In maize, genes that regulate anthocyanin production in the plant body include the regulatory loci, *B, Pl,* and certain alleles of *R,* and several genes for enzymes in the pathway, including *A1, A2, Bz1, Bz2,* and *C2.*

(b) Anthocyanins are dispensable for normal plant growth, so mutations in this pathway are viable.

(c) In appropriate genotypes, the entire mature shoot—including leaves, internodes, bracts, and anthers—produces anthocyanins.

(d) In some organs, anthocyanin production is tissue-specific. For instance, in maize, all tissues (including descendants of both L1 and L2 lineages) produce anthocyanins in the internode, while only the epidermis of the leaf blade is coloured. This is often advantageous because anthocyanin production in internal tissues can mask sectors in the epidermis; in maize, epidermal sectors lacking anthocyanin are more obvious in the leaf blade (where the underlying tissue is green) than in other parts of the shoot.

The disadvantages of anthocyanins include the fact that they are usually expressed late in the plant's life cycle, precluding their use as markers for AUT genes that are expressed early in plant development. Additionally, although multiple genes control anthocyanin production, this class of genes is not large (compared to photosynthetic mutations) and only certain regions of the genome can be marked using this approach.

3.3.3 Miscellaneous marker genes

Certain cellular traits, such as epicuticular wax, epidermal hairs, and cell shape and size are useful for marking cell lineages in the epidermis (9, 17). Furthermore, in cases in which an AUT gene is cell autonomous, the phenotype conferred by the AUT mutation itself may be used as a marker. For example, tassel branching was scored in the case of the *ramosa* mutation in maize (18); lack of carotenoids (white sectors in both the endosperm and in the shoot) was assayed in mutant *vp5* sectors in maize (19) and cell shape was observed in revertant *deficiens* sectors in *Antirrhinum majus* (20). Unfortunately, the inability to independently delimit the sector boundaries in these cases leaves the autonomy of these genes in doubt. It is possible, for example, that these genes are nonautonomous or are expressed tissue-specifically such that only sectors in certain locations, or of a certain size, are capable of expressing a mutant phenotype.

3.4 Scoring sectors

It is important to be aware of the problems associated with interpreting somatic sectors and to guard against them, if possible. In the case of sectors generated by ionizing radiation (Section 3.5.1), the presence of the marker phenotype in a sector does not guarantee loss of a linked AUT gene. Lesions limited to the M gene (e.g. interstitial deletions), as well as unlinked, second site mutations, would produce, for example, albino sectors without concomitant expression of the recessive AUT allele. Unless it is possible to use multiply marked chromosomes, the best control against these rare events is simply to generate a large number of sectors in order to define the most common type. In addition, hemizygosity of an unmarked locus (or loci) linked to an M mutation could have morphological effects independent of those induced by loss of the dominant AUT allele. For this reason, it is important to observe the phenotype of M sectors in plants that are homozygous for the wild-type allele of AUT locus. Finally, it is important to emphasize the difficulty of establishing the cell autonomy of genes that regulate multicellular traits (e.g. leaf shape), particularly when the structure in question is derived from several layers of the shoot apical meristem. In these cases the effect of a AUT sector may depend on the position of the sector within the structure, or the amount of tissue that the sector contributes to the structure.

3.5 Methods of generating sectors

3.5.1 Radiation-induced chromosome loss

Radiation-induced chromosome breakage leading to coordinated uncovering of a visible marker gene and the recessive allele of an AUT gene is the most commonly used method of clonal analysis. Among the factors that must be considered when using this approach is the radiation dose used to produce somatic sectors. This will vary in different situations and is a compromise between the need to use a dose high enough to produce a large number of sectors but low enough to limit nonspecific radiation damage. Radiation sensitivity is directly related to the genome size of a species, developmental stage, and in the case of seeds, to the degree of hydration (3). Doses ranging between 500 and 1500 R are effective for mitotic tissues of species with moderately large genomes, such as maize or tomato. On the other hand, *Arabidopsis thaliana* plants can tolerate 10–20 kR without suffering a significant amount of radiation damage. Dry seeds are significantly more resistant to ionizing radiation than hydrated tissue, so it is usually advisable to irradiate young seedlings, or seeds after imbibition.

Sector size is an important issue in the case of AUT genes that have a multicellular phenotype because it is often necessary to generate large sectors in order to observe the effects of these genes. This is particularly important if the affected structure is derived from several cell layers of the SAM as interactions between these layers may be critical in determining the trait of interest. Large sectors encompassing several cell layers of the shoot can be generated by irradiating embryos early in embryogenesis; unfortunately, due to the small number of cells in the embryo at this stage, the probability of obtaining sectors may be quite low. Alternatively, it is sometimes helpful to use genotypes that branch profusely since branches that fall within a sector may be derived almost entirely from that sector. The pattern of cell division within the shoot apical meristem during shoot growth and during the formation of lateral organs also has an important effect on the fate of somatic sectors. For example, in dicots, epidermal cells rarely contribute to internal tissue, whereas in some monocots epidermal cells frequently contribute to the internal tissue of the leaf and stem.

Finally, a word about radiation sources. Hard X-rays and γ-rays are almost equally effective in generating chromosomal rearrangements, although γ rays are more energetic and can, therefore, penetrate to deeper tissues. The choice between these types of ionizing radiation is usually based on pragmatic criteria such as the availability of one or the other source, and on the size of the material being irradiated. Most gamma sources can only be used to irradiate seeds or seedlings as the chamber of these machines is too small to accommodate larger specimens. X-ray sources of the type used for radiation therapy treatments (100–250 kV$_p$ machines) allow larger specimens to be irradiated. A filter (e.g. 2 mm Al) should be used to reduce the amount of low energy radiation produced. We have found that hospitals with radiation therapy facilities will often allow these sources to be used for the irradiation of plants.

The protocols given below refer specifically to maize, but we have used similar protocols for seeds and seedlings of tobacco, tomato, and cotton, and for inflorescences of tobacco and maize.

Protocol 2. Irradiation of imbibed maize seeds

1. Imbibe seeds by submerging them in 27°C water with vigorous aeration (e.g. an aquarium pump) for 30–35 h. Genotypes vary considerably in their rate of germination. Seeds are maximally sensitive when the radicle has just broken through the pericarp.

2. Spread imbibed seeds in a single layer in random orientation in a large Petri dish.

3. Irradiate with 1000 R of gamma rays (e.g. from a ^{137}Cs source) or hard X-rays (e.g. 250 kV$_p$, 15 mA, 2mm Al filter).

4. Keep seeds moist and aerated, and plant as soon as possible.

Lower leaves will have numerous small (< 1 mm) sectors, and may show signs of stress. Leaves six and above should have no apparent signs of radiation damage and should possess rare sectors expressing the recessive allele of the M gene.

Protocol 3. Irradiation of developing maize embryos

1. Plant seeds of the parental genotypes in large (e.g. 12 inch (30 cm) diameter) pots. (Unless a field irradiator is available, plants will have to be transported to the X-ray source when full grown.)

2. Pollinate ears to generate heterozygous seeds, and record date of pollination.

3. Irradiate well-pollinated ears 10–12 days after pollination with 500 R of hard X-rays. To ensure uniform treatment, irradiate each side of the ear with half the total dose (i.e. 250 R). If it is crucial to know the developmental stage of the embryo at the time of irradiation, several kernels should be harvested from each ear immediately after irradiation and fixed for histological observation.

4. Return plants to the greenhouse and allow ears to mature.

5. Score sectors on plants grown from kernels of irradiated ears.

Several examples of genes examined with this approach are worth mentioning. *Knotted1* is a semidominant mutation in maize that causes several morphological abnormalities in the leaf blade, including the production of bumps, or 'knots'. Radiation-induced sectors that uncovered an albino mutation (*lw*) linked to the wild-type *kn1* allele expressed a wild-type phenotype when these sectors encompassed all the tissue in the leaf blade, but did not eliminate the Knotted phenotype when they were confined to the epidermis or to the subepidermal mesophyll of the leaf blade (21, 22). Thus, the presence of *Kn1* tissue in the middle mesophyll of the leaf blade was sufficient to induce wild-type tissue (i.e. *lw kn1* tissue) in the epidermis to produce a mutant phenotype. The minimum requirements for expression of epidermal knots was the presence of *Kn* in half of a lateral vein and some inner mesophyll tissue. Sinha and Hake (21) concluded that *Kn1* does not diffuse laterally, but is noncell-autonomous in that vascular tissue in the middle mesophyll can induce epidermal cells to behave aberrantly. This nonautonomy of expression is particularly striking because the *knotted* gene contains a homeodomain, and is thus likely to encode a DNA-binding protein (23).

In contrast, the *Tp1* and *Tp2* mutations of maize appear to be completely noncell-autonomous. Both of these mutations are semidominant and have highly pleiotropic effects, including the transformation of reproductive structures into vegetative structures. Loss of *Tp1* was monitored by the expression of an albino mutation (*o5-1241*) of the proximal *O5* gene (24). Because no such marker existed for *Tp2*, this mutation was linked to the dominant wild-type allele of an albino locus (*Lw2*) on chromosome 5 by crossing *Tp2* on to a 5L–10L reciprocal translocation with breakpoints located distal to *Lw2* on 5L and proximal to *Tp2* on 10L; this placed *Tp2* distal to *Lw2*. This stock was then crossed to a *lw2*; *tp2* stock to generate *Lw2* T *Tp2*/*lw2*; *tp2* plants used for irradiation. Wild-type sectors of both *Tp1* and *Tp2* (i.e. albino sectors) had no visible effect on plant morphology (35), even when they occupied a large fraction of the shoot. This nonautonomous expression suggests that both of these genes may act through a diffusible signal.

The cell-autonomous expression of *Rp1* demonstrates yet another type of cellular interaction (or lack thereof). *Rp1* is a dominant gene that conditions resistance to the fungal pathogen, *Puccinia sorghi*. Bennetzen *et al.* (25) used the yellow green *oy* mutation to mark lineages that had lost the *Rp1* gene, and tested the autonomy of this locus by inoculating the pathogen on to leaves in *oy* sectors and in green tissues adjacent to sectors. Small, hypersensitive lesions typical of an *Rp1* response were observed in the green (*Rp1*) regions of the leaf, whereas large lesions characteristic of a susceptible *rp1* response were observed in the yellow-green (*oy rp1*) sectors. The fact that the hypersensitive response did not cross from green (*Rp1 Oy*) tissues into *oy rp1* sectors suggests that this gene not only controls the initiation of the hypersensitive response, but its propagation as well.

3.5.2 Transposition-induced chromosome breakage

An alternative method of causing chromosome breakage takes advantage of the transposable element system, *Activator (Ac)/Dissociation (Ds)*. Certain types of *Ds* elements (termed State I *Ds* elements) (26) and closely-linked Ac-Ds elements (27) cause chromosome breaks at their insertion site when they transpose. These elements are particularly useful for clonal analysis because they produce chromosome breaks at defined sites and therefore obviate the need for the M and AUT gene to be in a proximal–distal orientation. Moreover, they have the potential to be a generally useful method for clonal analysis since State 1 *Ds* constructs can be engineered to contain novel M genes suitable for clonal analysis and will induce chromosome breakage in heterologous species (J. English, personal communication). The following factors must be taken into account when using this method:

(a) A State I *Ds* element must be proximally located and linked in *cis* to dominant alleles of both the AUT gene and the M gene. The AUT gene and the M gene may be in any orientation relative to one another.

(b) In maize, *Ac*-induced transpositions of *Ds* usually occur late in somatic development, leading to mostly small sectors. This requires that the phenotype of the AUT gene be visible in small sectors, or that the experiment be conducted on a large enough scale to obtain rare, large sectors. The timing of *Ac* activity in heterologous systems may be quite different from that in maize, however.

(c) The timing of *Ac* activity (thus *Ds* transposition) is usually not under the control of the experimenter, although the potential for engineering inducible or tissue-specific *Ac* activity exists.

(d) In maize, *Ac* exhibits a 'negative dosage effect'. A single active copy of *Ac* generates larger sectors than multiple active copies.

(e) Chromosome breakage is limited to the *Ds*-containing chromosome. Thus, morphological artefacts of radiation damage are not a concern.

In maize, this method has been used to investigate the expression of a number of genes in the anthocyanin pathway (28, 29), a gene in the carotenoid pathway (*Vp5*; reference 19) and the morphological mutation, *Tp2* (35). *Ds*-mediated arm losses in *Ds Vp5/–vp5* plants produced well-defined white sectors (the *vp5* phenotype) in the endosperm and in vegetative tissue, demonstrating that this gene is cell-autonomous. A State I *Ds* element located proximal to *Tp2* on 10L was useful for the clonal analysis of this gene because the only good cell lineage marker on 10L (*W2*) is located at the distal end of the arm. *Ds*-induced chromosome breakage in *Ds Tp2 W2/–tp2 w2* plants caused coordinate loss of *Tp2* and *W2*, uncovering the *tp2* and *w2* (albino) loci. As previously noted for X-ray induced sectors of *tp2*, these sectors had a Teopod phenotype, demonstrating that *Tp2* is expressed noncell-autonomously.

3.5.3 Spontaneous chromosome loss

Chromosomes that are lost spontaneously during mitosis are another resource for clonal analysis. Two types of chromosomes that exhibit this behaviour are ring chromosomes and accessory 'B-A' translocation chromosomes. Ring chromosomes are lost during mitosis as a consequence of sister chromatid exchange, which results in the production of a dicentric chromosome (30). McClintock (30, 31) used ring chromosome loss to examine the somatic phenotype of several interstitial deficiencies in maize. However, this approach has not been widely used in maize because ring chromosomes are usually lost at such a high frequency during somatic development that they are difficult to maintain.

Accessory B-A chromosomes are easier to maintain than ring chromosomes because they are lost less frequently. Some of the advantages that these chromosomes offer for clonal analysis are shown below:

(a) In maize, B-A chromosomes covering a large fraction of the genome already exist, and have been genetically well characterized.

(b) Genes located on any part of the B-A chromosome may be used to mark the loss of the chromosome, thus obviating the need for the AUT gene and the M gene to be in a particular orientation.

(c) When a B-A chromosome is carried as a free duplication, loss of this chromosome produces euploid sectors rather than aneuploid sectors, which may have secondary deleterious effects.

(d) Because recombination occurs between the autosomal portion of the B-A chromosome and the normal autosome, dominant AUT or M alleles can be recombined on to B-A chromosomes if necessary.

(e) In some stocks B-A chromosomes are lost early in development, resulting in large somatic sectors.

The somatic behaviour of all the existing B-A chromosomes in maize has not been well characterized, so it is not clear if this approach is generally useful. Birchler (32) used this method to characterize the expression of genes affecting endosperm size, and we have used it to study the autonomy of *Tp1* (35). Because *Tp1* is dominant, it was necessary to recombine this mutation on to a B-A translocation. Recessive AUT mutations must be linked in *cis* to the recessive allele of an M gene present on the B-A translocation. Once this is accomplished, the stock carrying the recessive alleles of the AUT and M genes is crossed, as a pollen parent, to a stock carrying a B-A translocation with the dominant alleles of these genes. The B-A stock may be homozygous, heterozygous, or hyperploid for the translocation. F1 seeds or plants are then screened for sectors expressing the recessive allele of the M gene.

3.5.4 Reversion of transposon-induced mutations

Mutations caused by the insertion of a transposable element often produce wild-type somatic sectors when the transposable element excises from the gene during development. If the gene disrupted by the transposon has an obvious cellular phenotype, these revertant sectors fortuitously provide information about the cell autonomy of this gene. This is not the method of first choice for analysing cell autonomy because the lack of an independent cell lineage marker makes it impossible to determine the true extent of a sector. Consequently, the appearance of revertant sectors is not absolute proof that a gene is cell-autonomous. On the other hand, revertant sectors may demonstrate that a gene is expressed nonautonomously if the cells at the boundary of a sector express a phenotype intermediate between that of mutant and wild-type tissue.

Transposon-induced mutations of genes affecting anthocyanin and chlorophyll production in maize and other species provide numerous examples of both autonomous and nonautonomous expression. In maize, revertant sectors of the anthocyanin genes *A* and *Bz1* have a diffuse boundary indicative of nonautonomous expression, whereas revertant sectors of *R* and *C* have sharp boundaries indicative of cell autonomous expression. This behaviour is consistent with the fact that *R* and *C* encode transcription factors that regulate several genes in the pathway, whereas *A* and *Bz1* encode biosynthetic enzymes that produce diffusible intermediates (33). Transposon-induced mutations may also be useful for determining the cell-autonomy of morphological mutations. A transposon-induced mutation of *Antirrhinum majus, deficiens-621*, transforms petals into sepaloid structures (20). These sepaloid structures characteristically have small sectors of petal cells that presumably represent wild-type revertant sectors resulting from the excision of the transposon in the *def* gene. The fact that the sectors have a discrete boundary suggests that the *def* product is cell-autonomous, a conclusion that is supported by the observation that this gene encodes a product with homology to transcription factors (34).

4. Concluding remarks

Genetic mosaics are powerful tools for analysing gene function and cellular interactions in plants. The principal limitation to the use of this method is the difficulty in generating the appropriately marked mosaics. In classical genetic systems, such as maize, methods exist for circumventing many of the problems that may be encountered in producing mosaics for any particular gene of interest. Species in which these techniques do not exist will benefit in the future from the development of novel methods for marking and generating somatic clones, such as the introduction of marker genes and site-specific recombination systems from other organisms by transformation (see Golic,

Chapter 1). Novel dominant genes suitable for marking cell lineages may already exist in many different species in the form of the selectable or screenable markers (e.g. β-glucoronidase, streptomycin resistance) that are used when generating transgenic plants. However, whether or not it is possible to readily observe the loss of these traits in somatic sectors remains to be determined. Dominant negative genes are more useful as cell lineage markers because it is technically much easier to observe the presence of a trait in a sector (resulting from the loss of an inhibitory gene) than the absence of a trait. The ability to transform plants with a dominant negative gene that inhibits a visible cell autonomous trait (e.g. chlorophyll production) would make virtually any region of the genome accessible to clonal analysis.

References

1. Dawe, R. K. and Freeling, M. (1991). *Plant Journal*, **1**, 3–8.
2. Irish, V. F. (1991). *Curr. Opinions in Genet. and Devel.*, **1**, 169–73.
3. Poethig, R. S. (1987). *Am. J. Bot.*, **74**, 581–94.
4. Poethig, R. S. (1989). *Trends Genet.*, **5**, 273–7.
5. Stewart, R. N. (1978). In *The clonal basis of development* (ed. S. Subtelny and I. M. Sussex), pp. 131–60. Academic Press, New York.
6. Tilney-Bassett, R. A. E. (1986). *Plant chimeras*. Arnold, London.
7. Marcotrigiano, M. (1990). In *Biotechnology in agriculture and forestry*, Vol. 11. (ed. Y. P. S. Bajaj), pp. 85–111. Springer-Verlag, Berlin.
8. Turcotte, E. L. and Feaster, C. V. (1967). *J. Hered.*, **58**, 55–57.
9. Neilson-Jones, W. (1969). *Plant chimeras* (2nd edn). Methuen and Co., London.
10. Marcotrigiano, M. and Gouin, F. R. (1984). *Ann. Bot.*, **54**, 513–21.
11. Clayberg, C. D. (1975). *HortScience*, **10**, 13–15.
12. Szymkowiak, E. J. (1990). Interactions between cells derived from the three shoot apical meristem layers of tomato and related species in graft-generated chimeras (Dissertation). Yale University, New Haven.
13. Binding, H., Witt, D., Monzer, J., Mordhorst, G., and Kollmann, R. (1987). *Protoplasma*, **141**, 64–73.
14. Marcotrigiano, M. and Gouin, F. R. (1984). *Ann. Bot. (London)*, **54**, 503–11.
15. Carlson, P. S. and Chaleff, R. S. (1974). In *Genetic manipulations with plant materials* (ed. L. Ledoux), pp. 245–61. Plenum, New York.
16. Coe, E. H., Jr, Neuffer, M. G., and Hoisington, D. A. (1988). In *Corn and corn improvement* (ed. G. F. Sprague and J. W. Dudley), pp. 81–258. American Society of Agronomy, Madison.
17. Bossinger, G., Maddaloni, M., Motto, M., and Salamini, F. (1992). *Plant Journal*, **2**, 311–20.
18. Johri, M. M. and Coe, E. H., Jr. (1983). *Dev. Biol.*, **97**, 154–72.
19. Wurtzel, E. T. (1992). *J. Hered.*, **83**, 109–13.
20. Carpenter, R. and Coen, E. S. (1990). *Genes Dev.*, **4**, 1483–93.
21. Sinha, N. and Hake, S. (1990). *Dev. Biol.*, **141**, 203–10.
22. Hake, S. and Freeling, M. (1986). *Nature*, **320**, 621–3.
23. Vollbrecht, E., Veit, B., Sinha, N., and Hake, S. (1991). *Nature*, **350**, 241–3.

24. Poethig, S. (1988). *Nature,* **336,** 82–3
25. Bennetzen, J. L., Blevins, W. E., and Ellingboe, A. H. (1988). *Science,* **241,** 208–10.
26. McClintock, B. (1949). *Carnegie Inst. Wash. Yearbk,* **48,** 142–54.
27. Dooner, H. K. and Belachew, A. (1991). *Genetics,* **129,** 855–62.
28. Rhoades, M. M. (1952). *Am. Nat.,* **86,** 105–8.
29. Kermicle, J. L. (1978). In *Maize breeding and genetics* (ed. D. B. Walden), pp. 357–71. John Wiley and Sons, New York.
30. McClintock, B. (1938). *Genetics,* **23,** 315–76.
31. McClintock, B. (1940). *Genetics,* **25,** 542–71.
32. Birchler, J. A. (1980). *Genet. Res.,* **36,** 111–16.
33. Dooner, H. K. and Robbins, T. P. (1991). *Annu. Rev. Genet.,* **25,** 173–99.
34. Sommer, H., Beltran, J.-P., Huijner, P., Pape, H., Lonnig, W.-E., Saedler, H., and Schwarz-Sommer, Z. (1990). *EMBO J.,* **9,** 605–13.
35. Dudley, M. and Poethig, R. S. (1993). *Genetics,* **133,** 389–99.

4

Toxin ablation in *Drosophila*

HERMANN STELLER

1. Introduction

Methods for the selective removal of specific cells have been widely used to investigate the influence of a particular cell on the development of its neighbours. In the past, cell ablation techniques have primarily relied on physical methods, including classical surgical techniques that date back to the origin of developmental biology (1). More recently, laser microbeam ablation has been used extensively for developmental and functional studies in the nematode *Caenorhabditis elegans* (2, 3), and also to investigate certain aspects of insect development (4, 5). Unfortunately, the small size and opaque cuticle of *Drosophila* place severe restrictions on the widespread use of laser ablation in this organism. In addition, physical ablation methods are only practical for killing rather small numbers of cells, but fail when it is necessary to remove larger cell populations, in particular if its members are distributed widely throughout the animal. Finally, given the labour-intensive nature of the technique, it is usually not possible to produce a sufficient number of individuals needed for genetic, biochemical, or behavioural analyses. In principle, these limitations might be overcome by generating transgenic animals which express a toxin gene under the control of a cell-type specific promoter (6–9). This approach, termed toxin ablation, was first pioneered in the mouse. In its original form, this technique is more or less limited to the ablation of cells which are not required for the viability and/or fertility of the animal. In addition, because of the extreme toxicity of the bacterial toxins used for this purpose, namely diphtheria and ricin toxin, it is essential to avoid even low levels of transgene expression in other cell types. Unfortunately, this has proven to be extremely difficult for *Drosophila*. All attempts to use un-attenuated toxin genes have either failed or led to nonspecific damage, apparently due to low levels of ectopic toxin gene expression. In order to circumvent these shortcomings, several systems for the conditional expression of toxin genes have been developed (10–12). This chapter discusses the advantages and limitations of three different systems for toxin ablation in *Drosophila*.

1.1 Overview

Currently available toxin ablation techniques involve the following steps:

(a) Select the toxin gene system that is most appropriate under the circumstances, taking the particular advantages and shortcomings of the different systems into account (see Section 2).

(b) Select and characterize a suitable promoter that is capable of directing toxin gene expression in the cell type to be ablated.

(c) Construct a promoter–toxin fusion construct by placing the promoter fragment upstream of one of the cloned toxin modules, and insert the fusion gene into a *Drosophila* transformation vector (typically P-element based).

(d) Microinject the toxin transposon together with helper DNA into embryos and isolate transformants.

(e) Establish and characterize lines containing single transposon insertions ('single-insert lines')

(f) Verify the structure of the toxin transposon and, if possible, the expression of toxin RNA under nonpermissive conditions.

(g) Induce expression of the active toxin ('permissive conditions') and examine the successful and specific arrest of cellular functions in the targeted cells ('autonomous defects').

(h) Analyse the developmental and functional consequences of cell ablation.

2. Toxin genes

2.1 Diphtheria toxin

The diphtheria toxin (DT) is a member of the family of ADP-ribosylating bacterial enterotoxins, which also includes cholera toxin (13, 14). DT is produced by strains of *Corynebacterium diphtheriae* that are lysogenic for the bacteriophage β^{tox+}, and is secreted as a single polypeptide chain of 535 amino acids. The toxicity of this molecule is due to its highly potent inhibition of protein synthesis. Upon entry into a cell, DT catalyses the ADP ribosylation of the translation elongation factor 2 (EF-2). This causes a rapid block of protein synthesis and subsequent cell death. Due to the enzymatic action of DT, only minute amounts of toxin are required to kill a cell, and it has actually been claimed that the introduction of a single molecule can be sufficient for this purpose (15). The DT polypeptide contains two major functional domains: the A chain (amino terminal, 193 residues) containing the active site for ADP-ribosylation of EF-2, and the B chain (carboxyterminal, 342 residues), which allows the binding and entry of the toxin into cells. The A chain is sufficient for cell killing upon introduction into the cytosol, but since it cannot enter cells without the B chain, it is harmless in the extracellular space. These features of the DT have provided the basis for developing cell autonomous toxin ablation systems, and the A subunit without the signal peptide (DT-A) has been adapted for expression in eukaryotic cells. In addition, an attenuated mutant version of DT-A, tox-176, is available which has a drastically reduced level of toxicity (16). Finally, recent reports claim that DT possesses an endonuclease activity *in vitro* (17, 18), but the relevance of these observations to cell killing *in vivo* has not been firmly established.

In *Drosophila*, the use of the original DT-A toxin gene cassettes has been surprisingly difficult. In several instances, attempts to generate transformants carrying the DT-A gene driven by various tissue specific promoters failed, even though ablation of the targeted cells should have yielded viable animals. In one case, several transgenic flies expressing DT-A under the control of the promoter from the *ninaE* gene (which encodes the major rhodopsin, Rh1) were actually obtained, but these flies displayed various morphological defects which are not consistent with the highly specific expression of the *ninaE* gene (11). Apparently, some level of ectopic expression, possibly due to genomic enhancers near the site of insertion, is responsible for this problem. In order to achieve better control over the production of DT, and to extend toxin ablation to cells essential for the viability of the organism, conditional expression systems have been developed (10, 11). Their properties are further discussed below.

2.2 Ricin toxin

Ricin is a toxic lectin produced in the seeds of the castor bean *Ricinus communis*. It is derived from a nontoxic prepro-precursor protein that is converted to the active form by proteolytic cleavage (19). The mature protein consists of two polypeptides that are crosslinked via disulfide bonds. The A subunit (267 amino acids) is the actual toxin and contains *N*-glycosidase activity that specifically destroys the ribosomal 28S RNA, thus inhibiting protein synthesis. The B subunit (262 residues) contains a galactose-specific cell-surface binding and membrane translocation activity, permitting the entry of ricin into the cell. As in the case of DT, the A chain of ricin (ricin-A) cannot enter cells without the B subunit and remains in the extracellular space where it is essentially nontoxic. Therefore, the expression of recombinant ricin-A kills only those cells which synthesize the toxin, and the release of toxin by dying cells is not expected to damage the surrounding cells. A ricin A toxigene has been constructed and was originally used for ablation experiments in mice (8). Subsequently, a cold-sensitive mutation has been isolated and used for cell ablation experiments in *Drosophila* (12).

2.3 Conditional toxin mutants and their use for cell ablation in *Drosophila*

There are several obvious advantages of conditional toxin expression for cell ablation experiments. First, such a method would allow the recovery and maintenance of transgenic lines where a toxin can be expressed in cells that are essential for viability or fertility. Even if cell ablation resulted in an early developmental arrest, e.g. during embryogenesis, it would be possible to obtain sufficient material for detailed anatomical, biochemical, or even genetic studies. Secondly, the anticipated developmental expression pattern of a toxin construct could be verified experimentally by RNA *in situ* hybridization. Finally, a conditional system that offers good control over the timing of toxin production would permit the ablation of cells at a particular developmental stage, even in the absence of promoter elements with much broader ranges of expression. It would also allow the selective use of promoters that are expressed in different tissues during different stages of development. For example, a promoter that is expressed more generally early during embryogenesis and which has a second, cell type specific pattern at a later developmental stage would not be useful for cell-type specific ablation unless toxin expression could be suppressed during the initial phase of activity.

So far, three types of conditional toxin mutants have been obtained:

- a DT-A amber mutation (10)
- temperature sensitive DT-A mutations (11)
- cold-sensitive ricin A mutations (12)

From the currently available options, the amber system appears to produce the biggest difference between the 'on' and 'off' state of the toxin gene. It was designed with the rationale that a stop codon introduced near the amino terminus of the DT-A gene should essentially block the production of any toxin DT-A polypeptide. A sufficient level of suppression, and consequently toxicity, should be regained upon the introduction of an amber suppressor tRNA gene (20). An amber stop codon was introduced into the DT-A chain by *in vitro* mutagenesis, changing the tyrosine at amino acid position 28 (TAT) into an amber codon (TAG). The resulting amber mutant version of DT-A (DT-A^amb^) was inserted into the pUChsneo transformation vector (21) to produce pSK409 (10). This plasmid contains unique *Bam*H1, *Sal*1 and *Xba*1 sites upstream of the DT-A gene to accept promoters driving toxin expression (see *Figure 1*). The utility of this system was demonstrated by

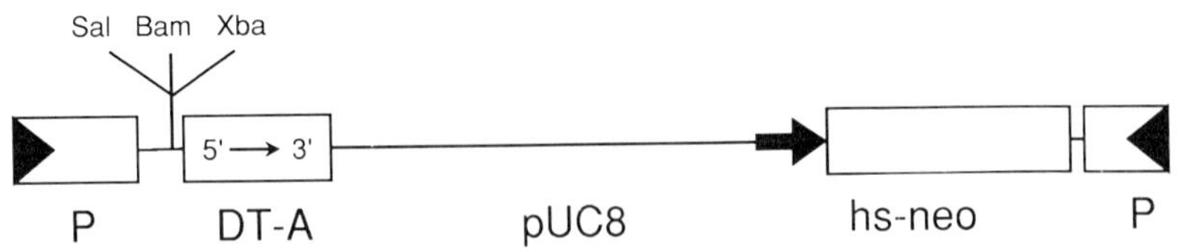

Figure 1. Map of pSK409. This plasmid contains an amber mutant version of DT-A inserted within the pUChsneo transformation vector (21). Three unique restriction sites (*Sal*1, *Bam*H1, and *Xba*1) are located immediately upstream of the DT-A gene and can be used to insert promoter fragments. For details, see reference 10.

using a photoreceptor cell specific promoter to drive expression of the DT-A^{amb} gene (10). In the absence of the tRNA suppressor, most photoreceptor cells differentiated and survived for extended time periods. In addition, toxin RNA accumulated in adult heads of the transformants, indicating that cells can tolerate mutant transcripts at least for several days. In contrast, the introduction of a Tyr-amber suppressor tRNA gene resulted in the complete ablation of all photoreceptor cells. The use of stage specific differentiation markers indicated that the differentiation of photoreceptor cells could be arrested within two days after the presumed onset of toxin gene expression. Nevertheless, even this system does not provide perfect control over the production of toxin, since some defects were observed in the absence of the tRNA suppressor. These limitations are discussed in more detail below.

Conditional toxin ablation systems that are based on temperature and cold sensitive mutations have also been developed. In principle, these systems should overcome many of the current limitations which arise from the lack of sufficiently specific promoters. Temperature and cold sensitive mutations of DT-A (11) and ricin A (12) were initially selected in yeast and subsequently expressed in the *Drosophila* eye. From this analysis, one temperature sensitive DT-A mutation, DT-A^{tsM} (11), and one cold sensitive ricin mutation, RAcs2 (12) emerged as the most useful toxin versions for cell ablation experiments in *Drosophila*. Both mutations produce only low toxicity at the nonpermissive temperature (28°C for DT-A^{tsM} and 18°C for RAcs2), and have significantly enhanced toxicity upon shifts to the permissive temperature (14–18°C for DT-A^{tsM} and 29°C for RAcs2). However, neither system succeeded in completely ablating all the targeted cells. Furthermore, the DT-A^{tsM} has no ADP-ribosylating activity, since it contains the same amino acid change as a previously characterized DT-A mutation. Therefore, it is possible that this mutation kills cells due to the aforementioned endonuclease activity that has been demonstrated for DT *in vitro* (17). Similarly, it is not clear whether the RAcs2 mutation has retained its 28S degrading activity or whether it exerts toxicity by some other mechanism. In both cases, incubation periods of at least 10 hours are required to produce extensive defects, and more complete ablation requires even longer temperature shifts. A comparative evaluation of both systems is very difficult at this point, primarily because different promoters were used for toxin expression. While the RAcs2 system produced more striking phenotypes, in particular upon shorter temperature shifts, it is likely that differences in the spatial and temporal activity of the two promoters that were used for toxin expression also contributed to this effect. The Rh1 promoter used to express the DT-A^{tsM} gene is activated at a very late stage in the differentiation of photoreceptor cells. In contrast, transcription of RAcs2 was induced at an extremely early stage in retinal development. Not surprisingly, toxin synthesis produces maximal effects before a cell has completely differentiated. Therefore, cells which are still dividing or are in an early phase of their differentiation programme are more severely affected by

Table 1. Choice of toxin system

Aim	System	(Reference)	Drawback
Complete ablation Tight control over toxin synthesis	DT-A^{amb}	(10)	need highly specific promoter two-component system
Temporal control Increase promoter specificity Convenience (one component system)	DT-A^{tsM} or RAcs2	(11) (12)	incomplete ablation unknown mode of action

blocking protein synthesis than fully differentiated cells. In order to achieve the most rapid and complete ablation possible under the circumstances, temperature shifts with DT-A^{tsM} or the RAcs2 constructs should be timed to induce toxin production at an early developmental stage. The considerations that influence the choice of the toxin system most appropriate for a specific purpose are summarized in *Table 1*.

3. Construction of promoter–toxin fusions

In order to ablate a particular cell type by toxin expression, it is essential to obtain a DNA fragment that is capable of specifically directing the expression of heterologous genes to this cell type. This obviously requires a fair amount of knowledge about the gene's expression pattern and its upstream control region. Given the limited ability to analyse the pattern of toxin RNA expression in transformants, it is essential to first test the activity of a promoter fragment using a fairly sensitive reporter system, such as *lacZ*. The pattern of *lacZ* expression should be examined for several different chromosomal insertion sites, since the presence of genomic enhancers in the vicinity of the transposon may influence promoter activity and lead to ectopic expression. This is of particular concern since the interpretation of toxin ablation experiments may become very difficult, if not impossible, in such a case. Therefore, it is essential to use a promoter fragment that is, as far as possible, immune to ectopic activation. Unfortunately, only little is known about the susceptibility of different promoters to genomic enhancers, and how to best shield promoters from ectopic activation (22). Unless information on suitable 'buffer sequences' is available, it may be possible to reduce the undesired effects of genomic enhancers by simply increasing the size of the promoter fragment. In addition, the promoter–toxin module should be flanked by other promoter elements on both sides, since these promoters may act as 'sinks' for any adjacent genomic enhancers. Fortunately, many P-element vectors, including pSK409 naturally provide for such an arrangement, since the toxin gene construct is placed between the P-element promoter (contained in the 5′-P element sequences) and that of a marker gene. If a DNA fragment produces the desired pattern of reporter gene expression in several different insertion sites, it can be linked to one of the cloned toxin modules. The pSK409 vector contains three unique restriction enzyme sites to allow the convenient introduction of promoter fragments upstream of the initiation codon (ATG) of the DT-A^{amb} toxin gene (see *Figure 1*). Once the proper construct has been made, the following steps (*Protocol 1*) complete the preparations for embryo injections.

Protocol 1. Purifying plasmid DNA for microinjection

1. Prepare plasmid DNA that is suitable for embryo injections. This usually involves double-banding in caesium chloride–ethidium bromide followed by Dowex AG50W-X8 chromatography (23). However, we have also had good success with a simpler protocol utilizing Qiagen columns (24).

2. Ethanol-precipitate the DNA and wash the pellet extensively with 70% ethanol to remove any salt.

3. Resuspend the DNA in injection buffer (5 mM KCl; 0.1 mM phospate, pH 7.8) at a final concentration of vector between 500 and 1000 μg/ml. Add pπ25.7wc helper DNA (25) from a concentrated stock solution in injection buffer (typically 2 mg/ml) to a final concentration of 100–150 μg/ml.

---

4. Germ-line transformation

The procedures for microinjection and germ-line transformation of *Drosophila* have previously been described in detail (23, 26), and these protocols should be consulted for the relevant technical details to generate transgenic flies carrying ablation constructs. Although other transformation systems have been developed recently, P-element-mediated gene transfer remains so far the most convenient and most widely-used system available for this purpose. When using the DT-A^{amb} system, it is advisable to select ry^{506} flies as the host strain for injections, since the suppressor tRNA transgene is marked with ry^+ (20). In this case, a *ry* background facilitates the identification of adult progeny containing the amber suppressor construct. The general scheme for the establishment of transgenic lines bearing single toxin gene inserts is outlined in *Figure 2*. Again the reader is referred to the excellent article by Spradling (23) for details of germ-line transformation and the establishment of transformed lines.

Since some investigators have encountered problems during the G418 selection of transgenic larvae, instructions for G418 selection are given below. G418 can be purchased under the name Geneticin as an impure preparation from Sigma or GIBCO. Because the actual G418 content in different batches

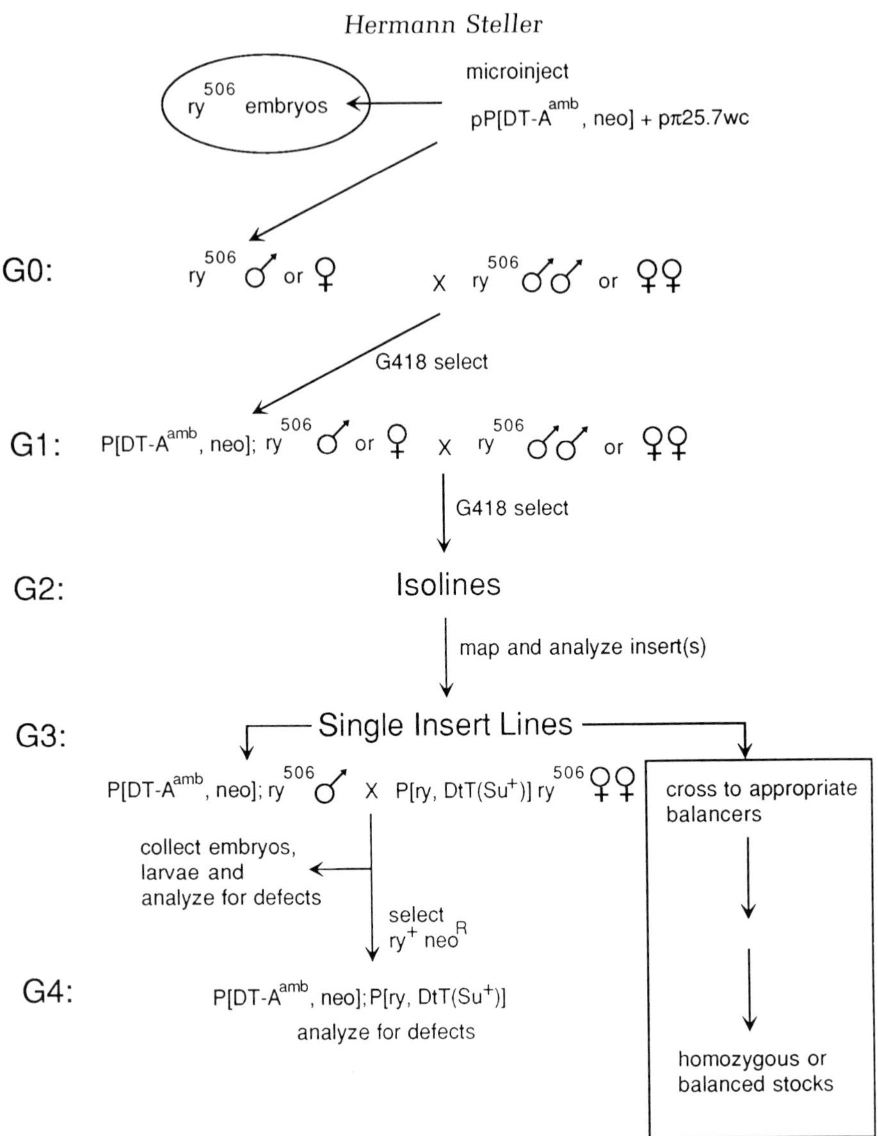

Figure 2. Scheme for using the DT-A^{amb} system to ablate cells in *Drosophila*.

may vary, some titration may be necessary to determine the optimal concentration in the food. A large number of G418-resistant transformants is available from *Drosophila* stock centres to provide a positive control during the titration process. Usually, a concentration of 500 µg/ml works well, but sometimes higher concentrations (up to 1000 µg/ml) may be necessary in order to avoid getting too many 'escapers'. If in doubt, it is better to accept a few escapers that can be eliminated in a second round of G418 selection, rather than losing transformants because of too stringent initial selection conditions.

Protocol 2. G418-selection of pSK409 transformants

1. Make a stock solution of 10 mg/ml of Geneticin in water and store at −20°C.

2. Prepare G418 medium by mixing 100 g dry yeast, 100 g glucose, 15 g agar and 30 ml Lexgard M stock solution (10% Lexgard M in ethanol, Inolex Chemical Company) with one litre of water. Boil for several minutes to ensure that all ethanol has evaporated, cool to 43–45°C and add G418 to 500 μg/ml final concentration (or desired amount). Pour into vials. Unused medium can be stored for up to two weeks at 4°C but must be properly wrapped to avoid dehydration.

3. Set up crosses between G0 flies derived from injected embryos and un-injected partners on normal food. Transfer to selective food after 2–3 days, or whenever egg-laying has fully started, and incubate at 25°C under humid conditions. For a small number of cultures, this can be easily accomplished by placing vials in a beaker which contains some water and is covered with aluminium foil. It is very important to avoid significant dehydration of the food, since this will otherwise affect the G418 concentration. Because G0 flies are not resistant to G418, they will suffer from prolonged exposure to the selective medium and eventually become sterile. Therefore, it is best to avoid exposing them to G418 until after mating has occurred.

4. Transfer flies on to new selective food every 2 or 3 days. It is absolutely critical to incubate all cultures at 25°C, because resistance levels are significantly lower at 18°C.

5. **Optional**: heat shock the developing culture upon removal of the G0 parents for 60 minutes at 37°C. Repeat this treatment once a day for 3 or 4 consecutive days. This heat-shock regime is intended to produce stronger expression of the neo^R gene, which is driven by the hsp70 heat-shock promoter. However, it is not essential for the recovery of transformants, since sufficient levels of resistance and transformants are also obtained without heat shocks. Many investigators find it difficult to remember to remove their flies after the appropriate time from the 37°C incubator and prefer to maintain their cultures constantly at 25°C.

6. Mate surviving G1 adults to appropriate partners and subject to another round of G418 selection.

---

5. Analysis of transformants

The strategies for establishing single-insert lines have previously been described in detail (23). Because of the possible influence of position effects on the expression of the transgene, it is critical to generate and analyse several transformed lines that contain a single transgene in different chromosomal insertion sites. If the transformation frequency is low, a serious concern when working with toxic gene constructs, additional insertion can be obtained by 'transposon jumping'. This is readily accomplished by using $\Delta 2$-3 as stable source of P element transposase (27). Conditional toxin systems allow the analysis of the expression of toxin RNA under nonpermissive conditions, although, so far, this has not routinely been done. For example, if transformants display unexpected phenotypes, one should be able to investigate the possibility of ectopic toxin expression with *in situ* hybridization techniques. In addition, the anticipated spatial and temporal expression of toxin RNA can often be verified by Northern analysis, provided that sufficient quantities of the relevant tissues and/or developmental stages can be obtained (10). Despite the lower resolution of this method compared to *in situ* hybridization, such an analysis can be useful for revealing ectopic toxin RNA expression.

The strategies for induction of toxin synthesis in transformed flies depend on the particular conditional system. In the DT-A^{amb} system, active toxin is produced by crossing transformants to flies carrying an amber suppressor tRNA (10, 20). Because the available Tyr-amber suppressor RNA strains are essentially male-sterile, male toxin gene transformants have to be crossed to females containing the suppressor RNA construct (see *Figure 2*). The presence of both the toxin gene and the suppressor tRNA can be confirmed in adult flies by selecting for both neo^{R} and ry^+ eye colour. However, such an assessment is more difficult at earlier developmental stages. In these cases, amber mutations in a reporter gene could be used (20). On the other hand, it is often possible to infer the presence of the tRNA suppressor directly from phenotypic analyses. In the other conditional systems, toxin production can be simply induced by temperature shifts (cold treatments or heat treatments, respectively). In order to ensure the most rapid and extensive ablation possible, it is best to perform the temperature shifts at an early stage in the differentiation of the targeted cell type. The animals should remain at the permissive temperature for at least 5–10 hours in order to induce significant defects.

Next, it is necessary to assess the specificity, extent, and speed of cell ablation in the targeted cell population. In this context, it is important to distinguish between an arrest of protein synthesis and actual ablation. Unlike laser ablation methods, toxin synthesis in a cell will not lead to its prompt destruction and removal from the organism. Depending on the type of cell and its state of differentiation, it may take many hours until a targeted cell disintegrates and physically disappears. Not surprisingly, dividing and un-differentiated cells appear to be more rapidly affected than fully differentiated cells. The strategies for assessing the extent and specificity of toxin-mediated cell ablation depend largely on the particular cell population and experimental circumstances. Generally, it is possible and necessary to investigate the autonomous and nonautonomous effects of toxin ablation simultaneously. This often involves the use of cell type specific markers, such as antibodies and *lacZ* enhancer detector lines (28). In particular, the use of an antibody corresponding to the gene whose promoter has been used to drive toxin expression allows a good assessment of how efficiently the targeted cell population is removed. If stage specific markers are available, they can be used to determine when cellular differentiation becomes arrested. The death and lysis of targeted cells are best examined using semithin plastic sections (light microscopy) or transmission electron microscopy. Staining with the vital dye acridine orange allows the convenient visualization of programmed cell deaths (29, 30). However, this dye is not suitable for the reliable detection of necrotic forms of death (30). Therefore, it should not be used to analyse cell death in the targeted cell population. On the other hand, it may serve as a convenient tool for examining whether programmed cell death is induced in other cells which may depend for their survival on interactions with the ablated cell type. Finally, genetic mosaic techniques can be used to induce toxin expression in only a subset of cells in the targeted population.

5.1 Preparation of tissue sections

The phenotypic analysis of toxin transformants can sometimes be adequately performed on whole mount preparations of embryos or dissected organs (31). However, more extensive analyses will often require the preparation of tissue sections for light (LM) or transmission electron microscope (EM) studies. The necessary technical details can be obtained from numerous articles and books that have appeared on this subject (31, 32). *Protocol 3* outlines the basic steps for tissue fixation, embedding, sectioning, and staining. With slight modifications, this protocol is applicable to many different developmental stages and tissues of *Drosophila*.

Protocol 3. Preparation of tissue sections for analysis

1. Follow either (a) or (b)
    (a) *Embryo preparations*
        Collect and dechorionate embryos in sodium hypochlorite (bleach) as previously described (31). Fix by shaking the embryos in equal volumes of heptane and fixative (2% glutaraldehyde, 1% paraformaldehyde, 1.5% acrolein in 0.1 M phosphate buffer, pH 7.0) for 20–30 minutes. Manually remove the vitelline membrane (31) and transfer to fresh fixative for another 30 minutes on ice.

Protocol 3. *Continued*

> (b) *Postembryonic tissues*
>
> Dissect larvae, pupae, or adults in either cold phosphate buffer or directly in fixative (avoid inhalation of vapours in this case). Remove any cuticle as far as possible and allow the tissue to remain in the fixative for one to several hours on ice.

2. Wash embryos or dissected organs in cold 0.1 M phosphate buffer (PB) and rinse twice. Transfer to 2% osmium tetroxide in PB and incubate for 1–3 hours on ice.

3. Wash in PB, followed by several changes of distilled water at room temperature.

   Optional: If preservation of membranes is of particular importance, the tissue can be treated with 1% uranyl acetate in water at 50°C for 12–16 hours. In this case, the tissue should be thoroughly rinsed in water after osmication to remove all traces of phosphate, since a precipitate will otherwise occur.

4. Dehydrate through an ethanol series (50%, 70%, 80%, 90%, 95%, and twice with 100% ethanol) for 10 minutes each. Transfer through two changes of propylene oxide (10 minutes each) into a mix of 50% propylene oxide and Spurr's epoxy medium (33) and allow to infiltrate for one hour.

5. Carefully remove the propylene oxide/Spurr's mixture and replace with pure Spurr's medium. Incubate the tissue for 6–16 hours at room temperature. Replace with fresh epoxy medium and transfer individual tissue pieces or embryos to embedding moulds. Polymerize the epoxy resin at 70°C for 24 hours.

6. Cut semithin (1–2 μm) or ultrathin sections (500–600 Å) for LM or EM analysis, respectively. EM sections may be stained with lead citrate and uranyl acetate (32). LM sections can be stained upon transfer to coverslips by adding a few drops of a solution of 0.1% toluidine blue in 0.1% sodium borate for 5 minutes at 55°C.

It is possible to obtain ultrathin sections adjacent to a semithin section, permitting an examination of the same cell at both the LM and EM level. Slight modifications of this protocol can also be used to obtain sections from whole mount preparations that have been stained for β-galactosidase activity (suitable for LM and low magnification EM work) or with antibodies (primarily LM work). In these cases, the tissue is washed in several changes of PB upon completion of the X-gal or DAB staining reactions and then refixed as described in step 1, *Protocol 3* (the acrolein can be omitted for LM preparations). If EM analysis is important, only very mild detergent conditions should be used during the initial staining protocol. In addition, toluidine staining can make the detection of X-gal difficult and is best avoided when analysing β-galactosidase activity at the LM level.

6. Interpretation of results

Since their inception, the interpretation of ablation experiments has often been difficult and has sometimes led to incorrect conclusions being drawn regarding developmental mechanisms. If a cell is damaged but retained in the tissue, the developmental consequences may be very different from what would have happened if the cell had been completely removed. This is a relevant concern for toxin ablation studies, since cells may retain their positions for at least several hours after protein synthesis has been blocked. Therefore, the fate of targeted cells should be carefully analysed as outlined in the previous section.

Another potential problem for interpreting the effects of toxin-mediated cell ablations on development is the possibility of ectopic toxin gene expression. Ideally, one may be able to diffuse such concerns by using *in situ* hybridization to carefully examine the pattern of toxin mRNA expression under non-permissive conditions. However, in reality it may be difficult to obtain sufficient sensitivity to eliminate the possibility that low levels of nonspecific toxin expression have occurred. Although very low levels of toxin may be insufficient for cell killing, they may suffice to interfere with the development and differentiation of cells. In *Drosophila*, ectopic promoter activity of transgenes is largely due to chromosomal position effects. The susceptibility of a promoter fragment to such ectopic activation should have been revealed with reporter gene systems during the initial promoter characterization (Section 3). Any promoter fragment that produces a variable pattern of activity in different chromosomal sites should be avoided. In addition, it should be obvious by now that a meaningful analysis of developmental defects cannot rely on a single transformed line.

These precautionary statements should not distract from the significant contributions that toxin ablation experiments may potentially make to reveal cell–cell interactions during development. Toxin ablation may often be the quickest and most straightforward method to initially examine cellular functions and cell–cell interactions during development. Interesting initial observations can subsequently be corroborated using other techniques. Furthermore, toxin ablation techniques may be useful for obtaining mutants that affect previously known cell–cell interaction phenomena. By isolating phenotypic suppressors of a conditional toxin phenotype it may be possible to recover interesting mutations, for example ones which affect the production of a developmental signal that is normally provided by the missing cell population. Finally, it is expected that an increased interest in toxin ablation techniques will lead to significant technological advancements during the next few years. Some of the current limitations and possibilities for future improvements are discussed below.

7. Current limitations and future prospects

Although three different strategies for toxin-mediated cell ablation in *Droso-phila* have been developed, they all have significant shortcomings. For one, all conditional systems are highly attenuated, even under permissive conditions. Consequently, cell ablation is not always as rapid or as complete as one may wish. Based on the efficiency of the amber tRNA suppressor in suppressing other mutations (20), one can estimate that the amount of toxin produced from the DT-A^{amb} gene under permissive conditions is 100–1000-fold lower than that of a wild-type gene. The relative levels of toxicity for DT-A^{tsM} and RAcs2 are completely unknown, and it is not even clear whether cell death in these cases results from the major enzymatic activities normally associated with these toxins (block of translation). Despite the high degree of attenuation, a small amount of active toxin is produced even under nonpermissive conditions in all these systems. Therefore, a significantly improved method that permits more rapid and complete cell ablation will have to satisfy two more or less opposing demands. On the one hand, to ensure rapid cell killings, such a system should permit the expression of relatively high toxin levels. On the other hand, toxin synthesis must be tightly restricted to the targeted cell population in order to avoid nonspecific damage. This may require improvements in both the cell-type specific expression of toxin genes (transcriptional control), and also the isolation of more stringent conditional toxin gene mutations (post-transcriptional control).

Another factor currently restricting the use of toxin ablation is the limited availability of well-characterized cell type specific promoters. Recently, a method for ectopic gene expression in *Drosophila* has been developed which may dramatically widen the applicability of toxin ablation (34). In this system, strains that express the transcriptional activator GAL4 in specific cells are generated by enhancer trapping. By placing a cloned gene under the control of a GAL4 regulatory element, its expression can now be selectively activated in GAL4-positive cells. This system has been used for the expression of DT-A and resulted in GAL4-dependent cell killing. Provided that the above outlined requirements for specificity are met, this methodology should allow the ablation of virtually any cell type in *Drosophila* without the need to extensively characterize and test promoter sequences.

Acknowledgements

This work was supported in part by NIH grant RO1-NS26451. I am grateful to Andrea Brand and Norbert Perrimon for communicating unpublished results, and to Hugo Bellen and Sam Kunes for comments on the manuscript.

References

1. Roux, W. (1988). *Virchows Arch. Pathol. Anat. Physiol.*, **114**, 113.
2. Sulston, J. and White, J. G. (1980). *Dev. Biol.*, **78**, 577.
3. Kimble, J. (1981). *Dev. Biol.*, **87**, 286.
4. Lohs-Schardin, M., Cremer, C., and Nüsslein-Volhard, C. (1979). *Dev. Biol.*, **73**, 239.
5. Doe, C. Q. and Goodman, C. S. (1985). *Dev. Biol.*, **111**, 206.
6. Palmiter, R. D., Behringer, R. R., Quaife, C. J., Maxwell, F., Maxwell, I. H., and Brinster, R. L. (1987). *Cell,* **50**, 435.
7. Breitman, M. L., Clapoff, S., Rossant, J,, Tsui, L.-C., Golde, L. M., Maxwell, I. H., and Bernstein, A. (1987). *Science,* **238**, 1563.
8. Landel, C. P., Zhao, J., Bok, D., and Evans, G. A. (1988). *Genes Dev.,* **2**, 1168.
9. Evans, G. A. (1989). *Genes Dev.,* **3**, 259.
10. Kunes, S. and Steller, H. (1991). *Genes Dev.,* **5**, 970.
11. Bellen, H. J., D'Evelyn, D., Harvey, M., and Elledge, S. J. (1992). *Development,* **114**, 787.
12. Moffat, K. G., Gould, H. H., Smith, H. K., and O'Kane, C. J. (1992). *Development,* **114**, 681.
13. Collier, R. J. (1975). *Bacteriol. Rev.,* **39**, 54.
14. Pappenheimer, A. M., Jr. (1977). *Annu. Rev. Biochem.,* **46**, 69.
15. Yamaizumi, M., Mekada, E., Uchida, T., and Okada, Y. (1978). *Cell,* **15**, 245.
16. Maxwell, I. H., Maxwell, F., and Glode, L. M. (1986). *Cancer Res.,* **46**, 4660.
17. Chang, M., Baldwin, R. L., Bruce, B., and Wisnieski, B. J. (1989). *Science,* **246**, 1165.
18. Bruce, B., Baldwin, R. L., Lessnick, S. L., and Wisnieski, B. J. (1990). *Proc. Natl Acad. Sci. USA,* **87**, 2995.
19. Olsnes, S. and Phil, A. (1982). In *Molecular actions of toxins and viruses* (ed. P. Cohen and S. Van Heyningen), pp. 51–105. Elsevier Biomedical Press, Amsterdam.
20. Laski, F. A., Gangulay, S., Sharp, P. A., RajBhandary, U. L., and Rubin, G. M. (1989). *Proc. Natl Acad. Sci. USA,* **86**, 6696.
21. Steller, H. and Pirrotta, V. (1985). *EMBO J.,* **4**, 167.
22. Kellum, R. and Schedl, P. (1991). *Cell,* **64**, 941.
23. Spradling, A. C. (1986). In Drosophila: *a practical approach* (ed. D. B. Roberts), pp. 175–97. IRL Press, Oxford.
24. *Qiagen plasmid handbook*, Spring 1992, pp. 12–21. Qiagen Inc., 9259 Eton Ave, Chatsworth, CA 91311, USA.
25. Karess, R. E. and Rubin, G. M. (1984). *Cell,* **38**, 135.
26. Santamaria, P. (1986). In Drosophila: *a practical approach* (ed. D. B. Roberts), pp. 159–73. IRL Press, Oxford.
27. Robertson, H. M., Preston, C. R., Phillis, R. W., Johnson-Schlitz, D. M., Benz, W. K., and Engels, W. R. (1988). *Genetics,* **118**, 461.
28. O'Kane, C. J. and Gehring, W. J. (1987). *Proc. Natl Acad. Sci. USA,* **84**, 123.
29. Spreij, T. E. (1971). *Netherlands J. Zool.,* **21**, 221.
30. Abrams, J., White, K., Fessler, L., and Steller, H. (1993). *Development,* **117**, 29–43.

31. Wieschaus, E. and Nüsslein-Volhard, C. (1986). In Drosophila: *a practical approach* (ed. D. B. Roberts), pp. 199–227. IRL Press, Oxford.
32. Osborne, M. P. (1980). In *Neuroanatomical techniques* (ed. N. J. Strausfeld and T. A. Miller), pp. 205–39. Springer, New York.
33. Spurr, A. R. (1969). *J. Ultrastruct. Res.,* **26,** 31–43.
34. Brand, A. and Perrimon, N. (1991). In Drosophila *Information Newsletter*, Vol. 1 (ed. C. Thummel and K. Matthews). Electronic mail publication, DIS-L@ IUBVM.UCS.INDIANA.EDU.

5

Cloning cell surface molecules by transient expression in mammalian cells

DAVID L. SIMMONS

1. Introduction

Since the mid 1980s, there has been a rapid increase in our knowledge about the specific cell surface molecules mediating cell–cell interactions and adhesion events. This has been largely due to the success of molecular cloning techniques, allowing the isolation of functional cDNA clones, encoding these glycoproteins.

There are a number of approaches that can be adopted for cloning cDNAs but one technique in particular, developed by Aruffo and Seed in 1987 (1–3), has been central to this enterprise. It is based on the transient expression of cDNA libraries in mammalian cells and the rescue of specific cDNA clones by antibody capture and panning. The efficacy of this procedure has transformed the field of clone isolation to such an extent that once a suitable antibody or ligand or cell line has been identified which recognizes a cell surface molecule, molecular cloning the cDNA encoding it is now an essentially trivial process.

Since 1987, a large number of cell surface molecules have been cloned using monoclonal antibodies to screen transiently expressed cDNA libraries including: the T-cell adhesin/activator CD2 (1) and its ligand LFA-3 (CD58) (3); the T-cell adhesin CD28 (2) recognizing B7/BB-1 (4); ICAM-1 (CD54) (5) and ICAM-2 (6) recognizing LFA-1 (CD11a/CD18); ELAM-1 (7) recognizing the carbohydrate ligand SLewisX (8, 9) (sialylated Lewis X blood group antigen); the B-cell marker CD22 (10) mediating adhesion of platelets and monocytes; CD44 (11–13) recognizing hyaluronic acid; Leu8/LAM-1 (14), the human homologue of the murine high endothelial homing receptor Mel-14, recognizing a carbohydrate ligand on endothelium; and CD36 (15, 16) an endothelial/myeloid protein which mediates the adhesion of *Plasmodium falciparum* infected erythrocytes to endothelium. VCAM-1, an endothelial adhesin for VLA-4 on lymphocytes, was cloned using a variation of the panning procedure employing cells directly as the recognition reagent (17, 18).

ICAM-2, an additional ligand for LFA-1, was cloned using the ligand itself (LFA-1) as a direct panning reagent (6).

This chapter describes the use of monoclonal antibodies for cloning cell surface molecules by transient expression of cDNA libraries and episomal rescue. Once cloned, the adhesion molecules can be functionally assessed by expression in COS cells and soluble forms produced from their extracellular domains allowing dissection of binding domains and quantitation of affinities and avidities.

1.1 Basic outline of transient expression and panning

The essential elements of this technique are outlined in *Figure 1*. It involves the construction of a representative cDNA library in a vector capable of replication and high-level expression in mammalian cells. After transfection of the library into the cell line, and the transient expression of surface molecule, a suspension of the cells is stained with the specific monoclonal antibody or ligand and then panned on plastic dishes coated with the appropriate second antibody. After washing, the panned cells are lysed *in situ* and low molecular mass episomal DNA is recovered by differential precipitation (Hirt procedure). Episomes are then transformed into *Escherichia coli* and plated. This cycle of transfection/transient expression/panning/rescue, usually needs to be repeated a further two or three times before individual recovered plasmids are analysed for expression of the specific cell surface molecule.

1.2 Advantages and disadvantages of transient expression cloning

The main advantages of transient expression cloning systems are:

(a) *Rapidity*. The transient expression profile reaches a maximum at 36–48 hours post-transfection. This means that each round of expression/ selection and rescue only takes 3 days, so a complete 3–4 round library screen can be completed within 2–3 weeks.

(b) *Full coding frame cDNA clones are isolated*. By definition, only those cDNAs encoding the entire reading frame of the protein will be cloned because the protein must at least have its ATG, extracellular domain, transmembrane domain, or lipid anchor and stop transfer sequence to give rise to a properly folded and processed surface molecule. In addition, as the selection is performed with monoclonal antibodies or direct ligands, the expressed molecule must substantially be the correct unmutated molecule.

(c) *Functional studies on cloned surface molecules*. The cloned cDNAs are in an efficient expression vector and can be used immediately for functional experiments, such as radioligand binding quantitation, cell adhesion studies, enzyme activity assay, etc.

5: Cloning cell surface molecules by transient expression

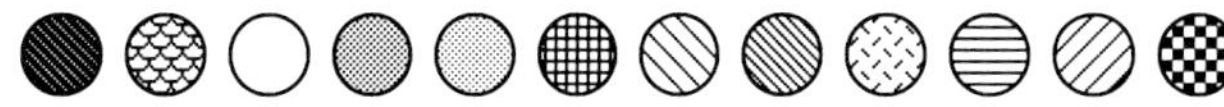

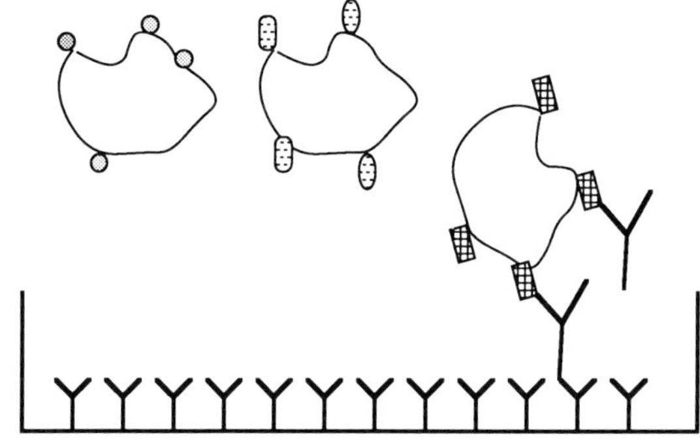

Figure 1. Outline of transient expression cloning.

For these reasons, cDNA based transient expression systems have become the method of choice for cloning cell surface molecules in recent years. *Table 1* lists the surface molecules cloned by such systems, up to July 1992, and any of the papers cited in this table provide useful references to the technique.

The major disadvantages of transient expression cloning systems are:

(a) *Multicomponent systems.* A major limitation of transient expression cloning systems is encountered by multicomponent glycoprotein complexes where the expression of any individual component of the complex at the surface of a cell requires the expression of all other members of that system. Clearly, only single molecules can be cloned by this system and such complexes would be missed. This is a major defect, since many of

95

Table 1. Cell surface molecules cloned by transient expression systems

Molecule	Distribution	Vector	Reference
CD1abc	T cell	pπH3M	1
CD2	T cell	pπH3M	2
LFA-3	pan haemo	pCDM8	3
ICAM-1	pan haemo	pπH3M	5
ICAM-2	pan haemo	pCDM8	6
CD7	T cell	pπH3M	29
CD19	B cell	pπH3M	30
CD20	B cell	pπH3M	31
CD22	B cell	pπH3M	10
CD28	T cell	pπH3M	2
LAM-1/Leu8	T cell	pCDM8	14
CD44H	pan haemo	pCDM8	11
CD44E	epi	mix	12
CD36	endo/my	mix	15
CD14	mono	pπH3M	32
CD31	endo/my	pCDM8	33
ELAM-1	endo	pπH3M	7
VCAM-1	endo	pCDM8	17
VCAM-2	endo	pCDM8	34
CD59	pan haemo	pCDM8	35
CD16	myeloid	pCDM8	36
CD32	B/myeloid	pπH3M	37
CD64	B/myeloid	pCDM8	38
CD34	myeloid	pCDM8	39
CD33	myeloid	pπH3M	40

Key. pan haemo, pan haemopoietic; endo, endothelium; my, myeloid; epi, epithelium; mono, monocyte; mix, pooled cDNA libraries in either pH3M or pCDM8.

the most important systems for cell recognition and signalling are multi-component complexes, for example, the T-cell receptor $\alpha\beta$ heterodimer requires expression of both chains for either chain in the heterodimer to reach the surface. The Ti/CD3 complex $\gamma\delta\epsilon\gamma\eta$ again requires multichain expression along with the TCR to achieve any surface expression of any of the CD3 chains. Integrins, major players in the process of cell–cell and cell–matrix adhesion, would also be missed by the expression cloning strategy, as these are $\alpha\beta$ heterodimers where expression of the α chain requires co-expression of the β chain for surface presentation. A way out of this cloning 'black hole', is the cotransfection of an existing expressing cDNA for one or all members of such complexes with the cDNA library under screen. For example, by cotransfection of integrin β chains with cDNA libraries, it is possible to clone α chains and vice versa.

It is also possible that the existing primate host cell integrins can act as 'surrogate mothers' for library derived α and β chains. Primate α chains could associate with human β chains giving rise to a species and chain heterodimer at the cell surface. Complementation of multichain complexes, by the use of host cell surface molecules may apply beyond the integrin family, but has not yet been tested.

(b) *Requirement for screening ligand.* A specific and high-affinity ligand is needed to screen the library. Usually, this has been a monoclonal antibody, although a variety of reagents can be used to identify cell surface molecules:

   i. monoclonal antibodies
   ii. labelled ligands
   iii. cells.

Direct use of receptor ligands, in a labelled form, have also been used, including FITC-labelled interleukin-6 (19), and iodinated GM-CSF (20).

Although the bias of this chapter is towards cell surface molecules, the transient expression system has been extensively used for the direct functional cloning of secreted molecules with biological activity, such as cytokines and growth factors. IL-3, IL-4, and GM-CSF, were cloned by this approach in the mid 1980s; supernatants from small pools of transfected COS cells were harvested and applied to appropriate bioassays, such as colony formation in soft agar, identification of positive pools, and repeat screening of positive pools. The cloning of all these cytokines led the way in transient expression technology. Cytokines are usually encoded by small mRNAs in high abundance so full-length cDNAs are likely to be well represented in cDNA libraries. In addition, the biological potency of many cytokines allows very small concentrations of active product to be detected in suitable mass assay systems.

Cell surface molecules do not satisfy either of these two convenient criteria. They are certainly not abundantly expressed genes; on average most cell surface molecules are present at 10 000–50 000 protein molecules per cell. Most surface molecules are over 1 kb and average round 1.5–2.5 kb. There is also the problem of a sensitive assay for detecting single clones in the library.

In the mid-80s a high-efficiency transient expression vector system (pCDM8) was developed in Brian Seed's laboratory at Massachusetts General Hospital, Boston. This allows the construction of representative cDNA libraries, high levels of expression, and accumulation of cell surface molecules ($1-5 \times 10^6$ molecules per cell surface). As CDM8 replicates to high copy number in COS cells, rescue and recovery of the episomal DNA is facilitated. Coupled with the design of a variety of ingenious screening systems, this system has greatly increased our knowledge about cell surface molecules by allowing the rapid cloning of cDNAs encoding them.

Clearly there are two parts to this technique:

- construction of a cDNA library
- expression and selection of that library

Both are described in detail in *Protocols 1–4 and 6*, respectively.

2. cDNA library construction and screening methods

2.1 cDNA synthesis and library construction

Protocol 1. RNA isolation

Based on the method shown in reference 21, amended by Brian Seed, Massachusetts General Hospital, Boston. This method allows increased capacity of cell/tissue masses to be used and also increases the speed of preparation (shorter centrifuge times).

A. *Total RNA isolation*

1. Dissolve (per ml cell lysate) 0.5 g guanidinium thiocyante (GuSCN) (Fluka) in 0.58 ml of 25% lithium chloride (LiCl). Filter through 0.45 μm filter. Add 20 μl of stock (14.4 M) β-mercaptoethanol.

2. Centrifuge cells (1000–1500 g, 5 min) in a 50 ml Falcon tube, alternatively use frozen tissue which has been ground into a powder in dry ice pellets. Disperse the pellet as a paste up the walls by tapping the side of the tube. Add 1 ml of the GuSCN/LiCl lysis buffer for up to 5×10^7 cells. Shear the lysate immediately in a polytron homogenizer, at top speed for 30–60 sec, until DNA viscosity has completely gone.

3. Follow (a) or (b)
   (a) *For small-scale preparations* ($< 10^8$ cells):
       layer up to 3.5 ml of the sheared lysate on to 1.5 ml of 5.7 M caesium chloride (CsCl) (RNAase free; 1.36 g CsCl added per 1 ml of 10 mM EDTA, pH 8.0) in a SW55 Beckman polyallomer centrifuge tube. Spin at 50 000 r.p.m., 240 000 g for 2 hours.
   (b) *For large-scale preparations* ($> 10^8$ cells):
       Layer 25 ml of the lysate on to 12.5 ml of the CsCl cushion in an SW28 Beckman polyallomer centrifuge tube. Spin at 24 000 r.p.m., 80 000 g for 8 hours.

4. Aspirate off the overlay through the CsCl interface and well down into the CsCl cushion, leaving only 1 ml in the bottom of the tube. Aspirate off all residual liquid from the walls of the tube and scour a ring just above the remaining liquid level. Invert the tube and cut the tube just at the rounded part. Wipe off any liquid with a Q-tip (cotton bud, cotton swab). Dissolve

the clear RNA pellet in 0.4 ml of RNAase-free water by triturating in a P1000 tip 10 times. Clear crystals of RNA should be visible that eventually dissolve.

5. Pipette aqueous RNA into an Eppendorf tube. Extract with phenol (0.5 ml), and then with chloroform (0.5 ml). Add 10% vol. of 3 M sodium acetate, and 2.5 volumes of ethanol. Place on dry ice for 10–15 min. Spin in a minifuge (12 000 r.p.m.) 13 000 g for 5 min. Decant the supernatant. Wash the pellet twice in 70% ethanol. Decant and remove residual ethanol with a P200 tip. Redissolve RNA pellet in 0.5 ml of RNAase-free water and titre (OD$_{260}$). Store RNA at $-70\,^{\circ}$C.

B. *Poly A+ RNA preparation*

Reagents

- oligodT cellulose (Collaborative Research type IV)
- loading buffer: 500 mM lithium chloride, 50 mM Tris (pH 8.0), 5 mM EDTA, 1% SDS
- middle wash buffer: 100 mM lithium chloride, 50 mM Tris (pH 8.0) 5 mM EDTA, 1% SDS
- RNAase-free sodium acetate
- RNAase-free water

Method

1. Prepare reagents by resuspending oligodT cellulose as 0.5 mg of dry powder per 1 ml of 0.1 M NaOH. Wash several times in RNAase-free water. Place in a plastic disposable 10 ml column, previously washed in 5 M NaOH and rinsed with water. Rinse oligodT in 2–3 column volumes of loading buffer (LB)

2. Pour the oligodT cellulose slurry into a sterile 15 ml Falcon tube and add 4–5 ml of LB, heat total RNA (1–2 mg at most) at 70$\,^{\circ}$C for 5 min. Remove from heat and chill on ice. Adjust concentration to 0.5 M with LiCl. Add the heat-treated RNA to the oligodT cellulose slurry and rotate on a wheel for 30 min at room temperature.

3. Decant into the disposable plastic column (Step 1) and wash with 5 volumes of LB. Then wash with 5 volumes of middle wash buffer.

4. Elute pA+ RNA with serial 0.4 ml fractions of RNAase-free water into Eppendorf tubes. Add 10% by volume of RNAase-free 3 M sodium acetate (pH 5.0), 2.5 vols of ethanol, and place on dry ice for 30 min. Spin for 10 min, 13 000 g, minifuge. Wash twice with room temperature 70% ethanol. Remove residual ethanol with a P200 tip and redissolve pellet in 100 μl of RNAase-free water. Peak fractions with 1–2 mg starting total RNA should contain 20–50 μg of reasonably pure polyA+ RNA. Fractions can be analysed on non-RNAase-free ultrathin 1% agarose minigels (see below).

Protocol 2. cDNA synthesis

Double-stranded cDNA is constructed by a simplified 'one-tube' version of the Gubler and Hoffman RNAaseH method (22), without NAD cofactors or any polishing of cDNA ends with T4 DNA polymerase.

A. *First strand*

Reagents
- RNAase-free water
- 5 × RT1 buffer: 250 mM Tris (pH 8.8), 250 mM KCl, 30 mM $MgCl_2$
- RNAase inhibitor: (Boehringer; 40 units/μl)
- oligodT: (dT12–18, Pharmacia)
- d × TPs: (dGTP, dTTP, dATP, dCTP: Pharmacia, Ultrapure)
- reverse transcriptase: the best but, unfortunately, the most expensive is Life Sciences RT-XL at 25 units/μl)

Method
1. Add 5 μg of mRNA to a sterile Eppendorf tube and heat to 100°C for 1 min. Quench on ice.
2. Adjust the volume to 70 μl with RNAase-free water.
3. Add the following:
   - 20 μl of 5 × RT1 buffer
   - 2 μl of RNAase inhibitor
   - 1 μl of 5 mg/ml oligodT
   - 2.5 μl of 25 mM dXTPs
   - 1 μl 1 M DTT
   - 2 μl reverse transcriptase
4. Incubate at 42°C for 40 min.
5. Heat inactivate at 70°C for 10 min.

B. *Second strand*

Reagents
- RNAase-free water
- RT2 buffer: 100 mM Tris (pH 7.5), 25 mM $MgCl_2$, 500 mM KCl, 50 mM DTT
- bovine serum albumin (BSA): molecular biology grade (Boehringer)
- DNA polymerase I: 5 units/μl (Boehringer)
- RNAase H: 2 units/μl (Boehringer)
- linear polyacrylamide carrier

Method

1. Add, to the same tube as in A:
   - 320 µl of RNAase-free water
   - 80 µl of RT2 buffer
   - 0.25 mg/ml BSA
   - 5 µl of DNA polymerase I
   - 2 µl RNAase H

2. Incubate at 15°C for 1 hour.

3. Switch tube to room temperature for a further hour.

4. Stop the reaction by adding 20 µl of 0.5 M EDTA, pH 8.0.

5. Extract with phenol: add 0.5 ml phenol, vortex, spin (13 000 *g* for 2 min in the minifuge), remove aqueous phase to a fresh tube.

6. Extract with 0.5 ml chloroform.

7. Precipitate by adding 10% vol. of 5 M NaCl, 5 µg linear polyacrylamide carrier, and adding 2 volumes of ethanol.

8. Place on dry ice pellets for 10 min.

9. Spin for 2–3 min only.

10. Wash twice with room temperature 70% ethanol, then remove residual ethanol with a P200 tip.

11. Redissolve cDNA pellet in 240 µl of water.

NB I have found that the addition of T4 DNA polymerase at the end of second-strand synthesis, to blunt the cDNA ends, does not appreciably or reliably increase the yield of ligatable cDNA (so it is simply omitted).

C. *Ligation of adaptors*

Reagents

- 10 × low-salt buffer: 60 mM Tris (pH 7.5), 60 mM $MgCl_2$, 50 mM NaCl, 2.5 mg/ml BSA, 70 mM-mercaptoethanol
- 10 × ligation additions: 1 mM ATP, 20 mM DTT, 10 mM spermidine, 1 mg/ml BSA, 100 mM $MgCl_2$
- BST XI adaptors: kinased by T4 PNK or prepared with 5′ phosphate at synthesis, see below
- T4 DNA ligase: 400 units/µl (New England Biolabs)
- LPA carrier
- TE: 10 mM Tris, 1 mM EDTA, pH 8.0

Protocol 2. *Continued*

Method

1. Add the following to the 240 μl of cDNA, from B, step 11:

   - 30 μl of 10 × low-salt buffer (10 × LSB)
   - 30 μl of 10 × ligation additions (10 × LA)
   - 5 μg (equimolar mixture of the Bst XI adaptors)
   - 1 μl of T4 DNA ligase
   - incubate overnight at 15°C

2. Phenol extract, chloroform extract, add NaCl (to 0.5 M) and 5 μg LPA carrier, then ethanol precipitate as above.

3. Resuspend cDNA in 100–200 μl of TE.

Protocol 3. cDNA size fractionation

The best way we have found for achieving the dual goals of efficient non-ligated adaptor removal and size fractionation of the cDNA, is kinetic density centrifugation on continuous gradients of 5–20% potassium acetate (KOAc).

1. Prepare continuous linear gradients in a 5 ml gradient maker.

2. Add 2.5 ml of 20% KOAc to the back chamber, and 2.5 ml of 5% KOAc to the front chamber.

3. Fill a Beckman 5 ml SW55 polyallomer centrifuge tube.

4. Layer the 100–200 μl of cDNA (*Protocol 2C*, step 3) very gently on to the top of the gradient.

5. Spin at 50 000 r.p.m. (240 000 *g*) for 3–4 hours at room temperature.

6. Puncture near to the bottom of the tube with a 21-gauge butterfly needle, collect 10 × 0.4 ml fractions.

7. Add 5 μg of LPA carrier, 2 volumes of ethanol, and freeze on dry ice for 10 min.

8. Spin for 3 min at 13 000 *g*, wash twice with room temperature 70% ethanol, and remove residual ethanol with a P200 tip.

9. Resuspend each fraction in 20 μl of water.

10. Analyse 2 μl of each fraction on a 1% agarose minigel, pool fractions[a] with cDNA larger than 500–750 bp.

[a] Fractions can be kept separate for each size band, e.g. 500–1000 bp, 1000–1500 bp, 1500–2000 bp, 2–3 kb, 4 kb or larger, resulting in 5 pools each with a tight size distribution.

Protocol 4. Ligation of cDNA to vector

A. *Small-scale test ligations*

1. Use 1–5% vol. of the cDNA

2. Ligate to a constant amount (10–20 ng) of vector (pCDM8, cut with BstXI and the stuffer fragment removed by KOAc gradient centrifugation as *Protocol 3*. Ligations are performed in a small volume (10–20 μl) with 10 ng of vector for 1 hour at room temperature.

3. Transform 10% of the ligation mix (1–2 μl) into 50 μl 'super competent' MC1061/p3 (see below).

4. Place on ice for 15 min, heat shock at 37°C for 5 min.

5. Plate on 10 cm LB plates containing ampicillin at 10 μg/ml and tetracycline at 10 μg/ml, with a 5 ml LB agar overlay poured during the heat-shock incubation to provide a 'drug-free zone' to enable the cells to grow and express drug resistance genes before being exposed to the antibiotics.

NB By using 1% of the cDNA and 10% of the ligation mix, the number of colonies per plate on this small scale is multiplied by 10^3 to give the maximum potential library size. A library size of 2×10^5–10^6 colonies should be aimed for. Alkaline/SDS miniprep DNA lysis should be used to analyse the inserts in at least 20–30 colonies. Use *Hind*III + *Pst*I or *XBa*I alone to liberate the inserts from pCDM8.

B. *Large-scale ligations*

1. Proceed to large-scale ligation using most if not all of the cDNA and proportionately more vector, if both criteria are satisfied (library size and insert size range). The entire cDNA yield should consume no more than 1–2 μg of the cut stuffer minus purified vector.

2. Ligate as above and transform into competent cells, ensuring that the ligation mix is kept at < 2–4% of competent cell volume.

3. Plate on 24 × 24 cm dishes at 10^5 colonies per plate.

4. Harvest the resulting primary plating and maxiprep using alkaline/SDS lysis and caesium chloride density gradients.

5. Store cDNA library as DNA at −20°C.

I have experienced no deterioration of library stocks over the 4 years I have been making them. Also, cDNA libraries can be safely amplified by re-transformation of the primary library stocks without gross loss of library complexity.

2.1.1 General methods

i. Ultrathin 1% agarose minigels

(a) Prepared on 'old-style' blood agglutination slides (Blue Star microslides, Proper Chance; 76 mm × 51 mm).

(b) 20–30 slides are made at once, layed out on Parafilm with fine tooth combs (6–7 wells per slide).

(c) Pipette 8–10 ml of warm (50°C) molten 1% agarose in TAE running buffer (Maniatis standard recipe). The agarose is held by surface tension as a bubble.

(d) These gels hold 10 µl/well and can be run extremely fast (200 V, 15 min) allowing rapid and easy monitoring of all the steps of the cDNA synthesis procedure obviating the use of tracer ^{32}[P]-label. They are extremely thin and have very low autofluorescence background allowing 10–50 ng of cDNA to be readily visualized by trans-UV illumination.

ii. Linear polyacrylamide (LPA) carrier

(a) LPA has proven to be a reliable and completely noninjurious inert carrier allowing efficient precipitation of picogram quantities of DNA at near zero cost.

(b) Prepared by polymerization of a 5% acrylamide solution with ammonium sulfate (0.1%) and TEMED (0.1%).

(c) No bis-acrylamide is present so only linear chains of polyacrylamide form. This solution is 50 mg/ml and a working dilution at 2 mg/ml is dilute. This is stored at −20°C and may be frozen and thawed many times. Usually, 5–10 µg per precipitation reaction is sufficient.

iii. Adaptor preparation

- Kinase adaptors are prepared enzymatically: adaptors at 1 mg/ml in 50 µl reaction volume, 5 µl of 10 × kinasing buffer (10 × KB = 0.5 M Tris (pH 7.5), 10 mM ATP, 20 mM DTT, 10 mM spermidine, 1 mg/ml BSA, 100 $MgCl_2$), 20 units of T4 polynucleotide kinase. Incubate at 37°C overnight.

- The non-self compatible BstXI adaptors are 5′-CTTTAGAGCACA-3′ and 5′-CTCTAAAG-3′.

- It is essential that the adaptors are efficiently HPLC purified before use, and that each new batch is tested on an existing batch of 'good' cDNA. I have found that good adaptors are one of the keys to good library construction.

iv. Super-competent cells

Many protocols exist for competizing cells. However, a simple two-step chemical method (originally derived from Mike Scott, Department of Neurology, UCSF), allows the routine production of cells with a competency of 1–5 $\times 10^8$. This level satisfies the dual need for cDNA library transformations and amplification of recovered episomal DNA from library screens (see below).

Protocol 5. Supercompetent cells

Reagents

- MC 1061/p3, *E. coli* strain
- TYM: 2% Bacto-tryptone, 0.5% yeast extract, 0.1 m NaCl, 10 mm $MgSO_4$
- transformation buffer I (TFBI): 30 mM KOAc, 50 mM $MnCl_2$, 100 mM KCl, 10 mM $CaCl_2$, 15% glycerol (v/v), do not adjust pH, sterilize by filtration
- transformation buffer II (TFBII): 10 mM Na–Mops (pH 7.0), 75 mM $CaCl_2$, 10 mM KCl, 15% glycerol

Method

1. Streak out MC1061/p3 on a fresh TYM plate. Incubate overnight at 37°C.
2. Pick single colonies into 5 ml of TYM, grow on a wheel with good aeration for 3–4 hours.
3. Dilute to 100 ml in a 250 ml flask with TYM, grow to midlog $OD_{600} = 0.5$.
4. Dilute to 500 ml in a 2 litre flask with TYM, grow to midlog $OD_{600} = 0.5$.
5. Rapidly chill cultures by swirling in water/ice mixture.
6. Pellet bugs in a Beckman J6 centrifuge in 1 litre pots at 4000 r.p.m. (2800 g) for 15 min.
7. Resuspend pellet very gently and slowly in 100 ml transformation buffer I on ice/water mixture.
8. Pellet at 2500 r.p.m. for 10 min at 4°C.
9. Resuspend pellet gently in 20 ml transformation buffer II at 4°C.
10. Aliquot in prechilled Eppendorf tubes and flash freeze in liquid nitrogen, store at −70°C.

NB Competency can be tested on supercoiled plasmid standard stocks at 100–10–1–0.1 pg levels, or by relative comparisons to existing tested batches of ligated cDNA or library screen Hirts. Competent cells maintain the desired level of competency for at least 3–6 months.

As with any library construction, the quality of the cDNA is of crucial importance to the isolation of any clones.

I have described the basic routine procedures used to construct cDNA libraries, even though a number of 'off the shelf' rapid procedures are now available from a variety of vendors that essentially allow one tube cells-to-pA$^+$ RNA preparation, usually involving oligodT derivatized magnetic beads as the affinity isolation method. These are very quick and reliable methods but suffer from being very expensive, especially if several libraries are to be made over a period of time. The same applies to cDNA synthesis kits. Many are now available, but if library construction is to be a routine part of laboratory practice the cost of such an approach would be prohibitive.

Table 2, describes all the cDNA libraries constructed in my laboratory (all freely available from me, or now distributed by the Human Genome Resource Centre, CRC, Watford Road, Harrow, London, HA1 3UJ), in the pCDM8 expression vector. A large number of cell surface molecules have been cloned from these libraries, and have been used by others for the isolation of many other internally expressed genes.

2.2 Details of vectors and expression systems

The choice of vector into which the cDNA is ligated, is linked to the choice of cell for expression. Molecular biologists have exploited mammalian DNA tumour viruses for this purpose. The two essential elements of these viruses are **origins** of replication, and ***trans***-acting DNA-binding proteins that interact with the origin and the polymerase/primase complex to replicate the viral genome to high copy number in the appropriate cell line. Two classes of virus have been exploited; papovaviruses, especially primate SV40 and murine polyoma, and ebnaviruses especially Epstein–Barr virus (EBV).

A crucial element of success in molecular cloning, is the copy number of the viral based vector in the host cell, for two reasons; first, it amplifies template per cell and thus increases the overall level of transcript production per cell; secondly, the amplified viral genome allows easy recovery from selected cells and thus reintroduction into *E. coli.*

This can best be illustrated by comparing vectors based on EBV versus SV40, as they differ markedly in their replication potential. EBV-based plasmids (23, 24), usually carry both the origin of replication (ori P) and the *trans*-acting origin amplifier (EBNA1), and can thus be expressed and amplified in any cell type. SV40-based plasmids usually only carry the origin of replication and have to be introduced into cell lines containing integrated copies of crippled SV40 genomes expressing the SV40 replicator protein large T antigen.

EBV-based plasmids only replicate to low copy number, typically 1–10 copies per cell, giving adequate levels of expression of specific molecules but posing difficult problems for the subsequent recovery of those plasmids from selected cells. Indeed, higher levels of EBV episomes per cell are often toxic to the cell and cannot be tolerated. Typically, recovery of EBV episomes has had to utilize the very high efficiency of phage lambda packaging extracts in order to rescue the vectors.

Table 2. cDNA libraries constructed in pπH3M or pCDM8

Human

1	HPBALL (peripheral blood, acute lymphocytic leukaemia)
2	JY (lymphoblastoid, B EBV +ve)
3	HepG2 (hepatocellular carcinoma)
4	U937 (promonocytic leukaemia)
5	U937/PMA stimulated
6	K562 (erythroleukaemia)
7	K562 (haemin stimulated)
8	LAK (lymphokine activated killer cells)
9	YT (HTLV-I +ve adult leukaemia, T cell)
10	HL60 (promyelocytic leukaemia)
11	HL60/g interferon stimulated
12	HT1080 (fibrosarcoma)
13	G361 ⎫
14	C32 ⎬ amelanotic melanomas
15	Placenta—full-term, normal pregnancy
16	Placental trophoblast (sorted 1st trimester)
17	Placental villi (1st trimester)
18	Human bone marrow (aspirate, ALL +ve, 1st remission)
19	HEL (human erythroleukaemia)
20	HUVEC (umbilical vein endothelial cell line)
21	HUVEC/stimulated with IL1-β (4 h)
22	HUVEC/stimulated with HT29 conditioned medium (48 h)
23	HUVEC/stimulated with DX3 conditioned medium (48 h)
24	L920 Hodgkin's lymphoma line
25	Fetal brain, 15–16 weeks
26	Normal colon
27	Colon carcinoma (solid tumour)
28	HT29 (colon carcinoma)
29	KG1 ⎫
30	KG1A ⎬ myeloblastic leukaemic lines
31	KG1B ⎭
32	K562 haemin stimulated
33	SU-DH-LI diffuse histiocytic lymphoma (non-Hodgkin's lymphoma)
34	Mel DS1 amelanotic melanoma CD36+
35	Mel DS1 amelanotic melanoma-X ray induced
36	Eosinophil
37	Fetal Muscle
38	Natural Killer Cell
39	CEM (T cell)
40	Tonsil
41	HU-PC (phaeochromocytoma)
42	LAD (leukocyte adhesin deficient patient EBV-B cells)
43	Normal human B cells (EBV transformed)
44	DX3 melanoma
45	HCT116 colon carcinoma

Rodent

1	Mouse B cells (LPS)
2	Mouse T cells (ConA)
3	Mouse thymocytes
4	IC21 mouse macrophage cell line, PMA stimulated
5	Mouse spleen (NOD mouse)
6	Mouse bone marrow aspirate
7	Rat alveolar macrophage/γIFN stimulated
8	Mouse serum stimulated macrophages

In contrast, SV40-based plasmids replicate to very high copy numbers per cell, typically 10^3–10^5, yielding very high expression of specific molecules and also allowing relatively easy recovery of the episomes from the selected cell. SV40 replicons are a burden to the cell in the long term, and cells bearing them have elevated morbidity and eventual mortality. However, the burden can be supported for a sufficiently long time to allow expression selection and recovery of those cells.

For this reason, papovavirus-based plasmids have been the most widely used system for transient expression and rescue. Of the many types of papovaviruses, SV40 has been used most frequently, though the murine permissive virus, polyoma, has also been exploited.

There are many variants of SV40-based plasmids, and only a few will be described here. All share the same basic features; the SV40 origin of replication (a 350 bp fragment of the SV40 genome); a eukaryotic enhancer and promoter driving high-level expression of the inserted cDNA or genomic fragment; downstream transcript processing elements (usually an intron and polyadenylation signal); and finally a prokaryotic origin of replication and some system for drug selection in *E. coli*.

The plasmids pH3M and pCDM8, developed by Aruffo and Seed in 1987 (1–3), have been successfully employed for the cloning of numerous cell surface molecules, pCDM8 postdates pH3M, and has now superseded it. Consequently, pCDM8 will be described in detail.

Figure 2(a) illustrates the pCDM8 vector. It contains the powerful cytomegalovirus (CMV) enhancer and promoter driving expression of cDNA inserted at a polylinker cloning site flanked by nonpalindromic BstXI sites. Downstream of this site is an intron and polyadenylation site, allowing efficient transcript processing and transport. pCDM8 contains both an SV40 origin of replication and a polyoma origin allowing replication of this vector in either primate cell lines such as COS-1 and COS-7 cells, and also murine polyoma transformed lines such as WOP and COP. This is particularly useful if a specific monoclonal antibody crossreacts with glycoproteins on the surface of COS cells, which are after all high primate cells of fibroblast/epithelial origin. Monoclonal antibodies raised in mice are highly unlikely to react with the surface of murine cells. The remaining elements of the vector allow replication in *E. coli* and drug selection mediated by a suppressor tRNA (supF) which suppresses amber stop codons in ampicillin- and tetracycline-resistance genes carried on a stable episome, p3, in the strain MC1061/p3. There is an M13 origin of replication allowing production of single-stranded templates of the plasmid when appropriate *E. coli* strains are superinfected with filamentous f1 phage, such as M13. A T4 DNA promoter is included at the 5′ edge of the cloning site to allow *in vitro* production of RNA templates for transcript terminus mapping and transcript production in cells permitting expression of T4 RNA polymerase.

An alternative version of pCDM8, designated pKS-1 (*Figure 2(b)*, G.

Stark, ICRF, personal communication), has been constructed in which the M13–SV40 origin portion of pCDM8 has been deleted and replaced with a pUC origin and ampicillin-resistance gene and SV40 fragment. This allows selection of recovered plasmid in any highly competent *E. coli* strain capable of ColE1 replication, and is not confined to the MC1061/p3 system.

A more radical variant pJFE14, constructed by John Elliott (25) (*Figure 2(c)*), utilizes the SRα promoter and R and U 5' regions of the human T-cell lymphotropic virus I. An intron from the 16S RNA is placed upstream of the BstXI polylinker cloning site. The plasmid contains a colE1 replicon and ampicillin resistance. Proponents of this vector argue that it overcomes some of the plasmid instability observed in the pCDM8–MC1061/p3 system.

All these vectors only replicate to high copy number in cells bearing SV40 or polyoma genomes. In 1981, Gluzman (26) produced SV40-transformed African green monkey kidney fibroblasts (CV-1) cells bearing integrated copies of the SV40 genome, crippled by the deletion of several bases at the SV40 origin. The resulting cells, COS-1 and COS-7, express high levels of SV40 large T antigen and permissivity factors, allowing high levels of SV40 replication per cell, but do not produce infectious viral genomes so are safe to work with under low-level containment. It is of considerable importance that these cells are eminently transfectable; with a DEAE–dextran/chloroquine regime (see below and references 27, 28), it is routinely possible to achieve 50–60% of total transfected cells expressing the introduced product.

A future development would be the construction of SV40-based plasmids which also produce SV40 large T antigen similar to the oriP/EBNA-1 p201–p205 system. Any cell line could then be transfected, irrespective of whether it contained endogenous SV40 genomes, and would allow genetic defects in defined cell lines to be complemented by introduced libraries, and allow rescue of the complementing episome.

There are many single genetic defects in cells which could be complemented by such a system, including those in DNA repair enzymes (xeroderma pigmentosum, ataxia, telangiectasia, Bloom syndrome, Fanconi anaemia), glycosylation enzymes, etc.

2.3 Screening by transient expression, panning, and rescue

First, I shall describe the principles involved in screening. The experimental details are given in *Protocol 6*.

cDNA libraries constructed as above in the pCDM8 vector are transfected into COS cells using DEAE–dextran as a facilitator (27, 28) and chloroquine diphosphate to reduce lysosomal degradation of endocytosed DNA (*Protocol 6A*). Forty-eight to seventy-two hours post-transfection, cells are lifted and incubated with Mabs as tissue culture supernatants at minimal dilution (1/10 at most), washed and applied to bacterial Petri dishes precoated with affinity-purified goat anti-mouse IgG.

Cells are allowed to 'pan' for 2–3 hours at room temperature and plates are

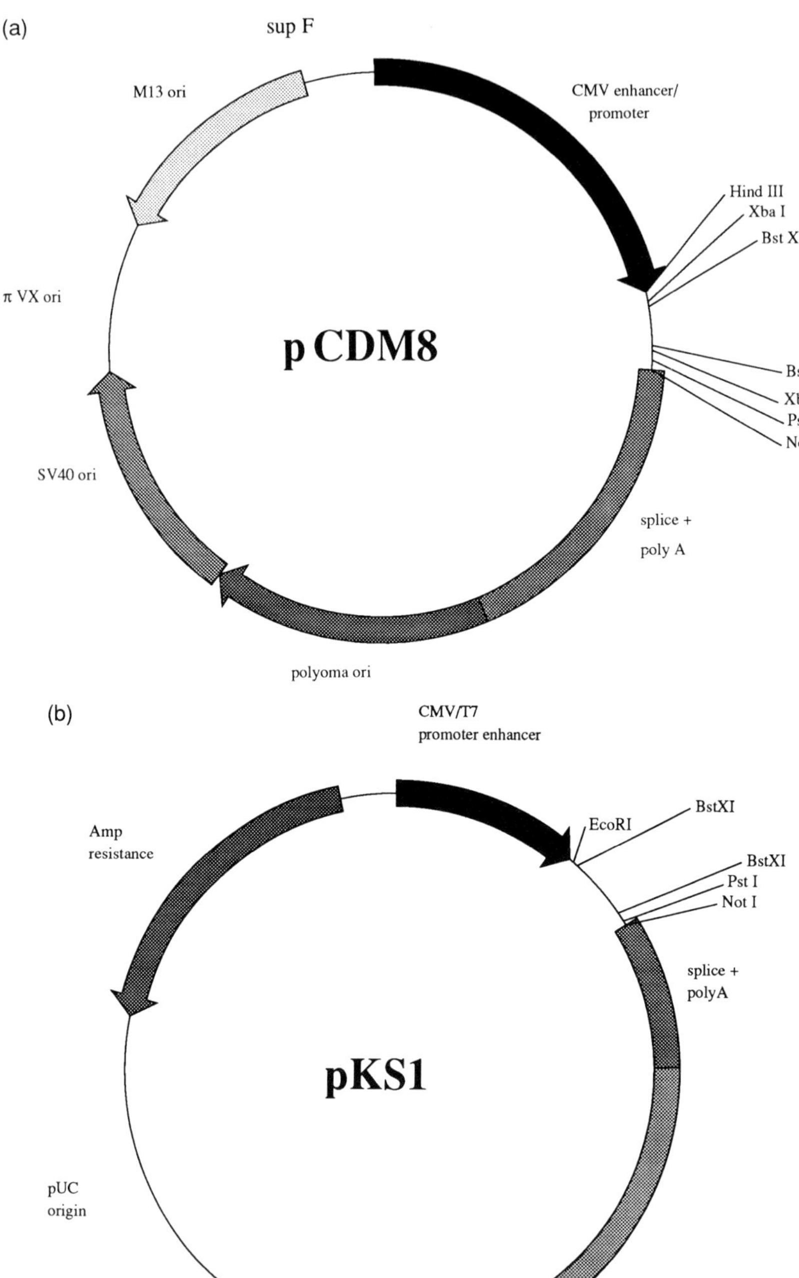
(a)
sup F
M13 ori
CMV enhancer/
promoter
Hind III
Xba I
Bst XI
p CDM8
Bst XI
Xba1
Pst I
Not I
π VX ori
splice +
poly A
SV40 ori
polyoma ori
(b)
CMV/T7
promoter enhancer
Amp
resistance
EcoRI
BstXI
BstXI
Pst I
Not I
splice +
polyA
pKS1
pUC
origin
polyoma
origin
SV40 origin

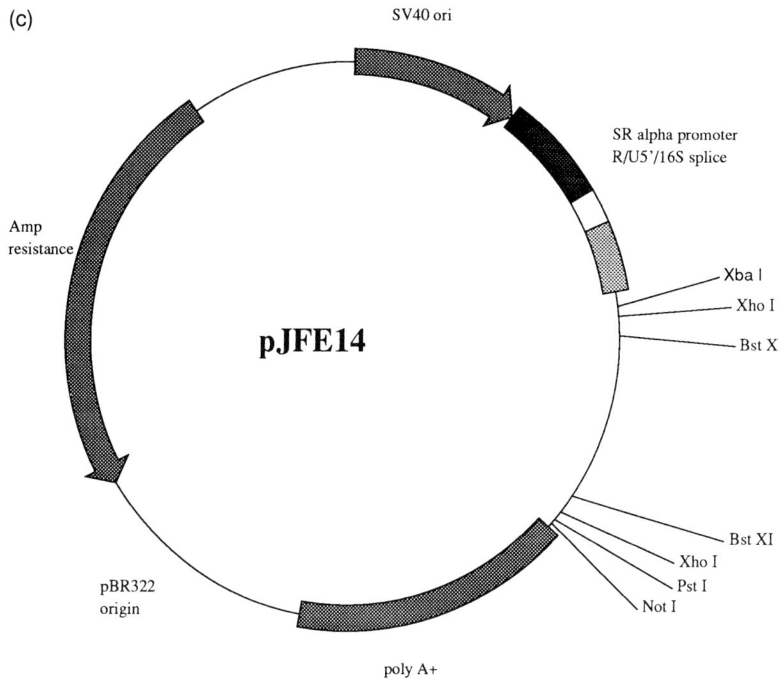

Figure 2. Eukaryotic expression vectors used for transient expression cloning. (a) pCDM8; (b) pKS1; (c) pJFE14. CMV, cytomegalovirus enhancer/promoter; sup F, suppressor tRNA; ori, origin of replication.

then washed gently three to five times. It is possible to observe panned cells (10–100) per dish even at the first round of selection. However, on many occasions no cells may be seen if the clone being sought is of very low frequency in the library. Whether panned cells are observed or not at this stage, the procedure must be continued for a further two rounds before a definitive assessment of success or failure is made. The panned cells are lysed *in situ* by applying 'Hirt squirt', the high molecular mass primate genomic DNA and episomal DNA are recovered according to *Protocol 6D*. A fraction of the recovered episomal DNA is transformed into highly competent MC1061/p3 and plated on LB agar containing ampicillin and tetracycline. A yield of 10^3–10^4 bacterial colonies should be obtained at this stage.

It is possible to continue to introduce the selected cDNA population into COS cells by DEAE–dextran facilitated transfection, however, a change of entry method is needed at this point. DEAE–dextran is a very efficient method for introducing DNA into cells, so it is an ideal method for the first round of screening to maximize representation of the cDNA library in COS cells. It is estimated that up to 10^3–10^4 different cDNA clones may be taken up by each COS cell by this method.

Consequently, a panned cell expressing the clone of interest will also contain 10^3–10^4 irrelevant cDNA clones, which will be represented in the yield of bacterial colonies from the first round. Thus, if DEAE–dextran was used for all subsequent rounds, a plateau of enrichment would be reached where the cDNA clone of interest would be contained within a heterogeneous population of irrelevant clones. This would mean that a very large number of individual bacterial colonies would have to be analysed by miniprep DNA isolation, individual transfection, and Mab staining.

To prevent this, the second round of screening (*Protocol 7*) is initiated by introducing the bacterial colonies into COS cells as protoplasts. This is a very inefficient technique; (1–5%) of COS cells are transfected, a small number of protoplasts actually fuse with each COS cell, and each protoplast obviously contains only one cDNA clone.

This means that a much smaller population of cDNA clones is introduced into each COS cell. The complexity of the resulting second round Hirt is thus greatly reduced and enrichment for the clone of interest is greatly enhanced.

The bacterial population is grown in liquid culture, plasmids are amplified in the presence of spectinomycin, converted to protoplasts by osmotic shock, EDTA chelation, and lysozymal digestion. Protoplasts are introduced into COS cells by polyethylene glycol-mediated membrane fusion. After another 36–48 hours to allow transient expression of plasmid encoded products, the COS cells are again incubated with the Mab, washed, and panned. As before, very few cells may be observed by visual scanning of the panning plates; although considerable enrichment has occurred as a result of the first round of selection, the switch to a more inefficient method of transfection means that a similar number of COS cells will pan at this round. A Hirt preparation is made and processed in the same way. Again, 10^3–10^4 colonies should derive from this round. A further round of protoplast fusion is needed before a definitive assessment of the success of the screening can be made (*Protocol 8*). Cells should be visible on the final third round panning dish. Individual bacterial clones are picked, DNA isolated by standard minipreparation methods and a fraction of that DNA is transfected into COS cells by DEAE–dextran facilitation. Forty-eight hours later the COS cells are stained *in situ* with the Mab, stained with a goat-anti mouse FITC second antibody, and scored by indirect fluorescence microscopy. Results at this stage are never doubtful. Only a small number of individual colonies need be analysed at this stage since 10–100% of these clones should be the clones of interest.

Protocol 6. Screening cDNA libraries by transient expression, panning, and episomal rescue—*Round 1*

This protocol describes DEAE–dextran transfection, expression, screening, panning, episome rescue, and transformation.

A. *Transfection*

1. Grow COS cells in DME/10% calf serum at 50–75% confluency (most conveniently in Falcon 15 cm intergrid culture dishes).

2. Use 10–20 μg cDNA library in pCDM8, or similar SV40-based vector, for 1×10^7 COS cells using 400 mg/ml DEAE–dextran (Sigma M_r 400 000) as a facilitator (27) and chloroquine diphosphate at 100 mM. Dilute cDNA library DNA well below 1 mg/ml in TE (*Protocol 1*), add DEAE–dextran (27, 28) and dilute up in medium either without serum or with 10% NuSerum (Collaborative Research) or a low protein concentration serum supplement such as UltroserG from Gibco–BRL.

3. Leave on for up to 4 hours, or until the COS cells begin to look vacuolated.

4. Aspirate medium, add 15 ml PBS/10% dimethyl sulfoxide (DMSO), osmotic shock medium for 2 min.

5. Aspirate and replace with regular medium.

6. Trypsinize cells, 24 hours post-transfection, and replate on fresh culture dishes to remove residual adsorbed DEAE–dextran. It is essential to do this in order to be able to lift the COS cells with EDTA the following day and achieve a monodisperse cell suspension.

B. *Screening*

1. Lift cells with phosphate-buffered saline containing 2 mM EDTA 48–72 hours post-transfection.

2. Wash in 'panning buffer' (PB) (PB = PBS (pH 7.2), 2 mM EDTA, 0.02% sodium azide, 5% fetal bovine serum) at 4°C.

3. Incubate with Mabs (either as tissue culture supernatants (1/10 at most) or ascites or purified antibody), at 4°C for 30 min.

4. Wash twice in cold PB.

C. *Panning*

1. Prepare panning plates:
- Prepare panning plates by coating 10 cm Falcon Petri dishes with 5 ml of a 10 μg/ml solution of affinity-isolated goat anti-mouse IgG in 50 mM Tris, pH 9.5, for 1–2 hours.
- Wash 3 times in PBS.
- Block remaining sites by overnight incubation with 5 ml/dish of blocking buffer (PBS (pH 7.2), 2 mg/ml BSA).
- Aspirate blocking buffer, and store plates at −20°C for up to 6 months.

2. Apply cells from (B) to room temperature panning plates.

3. Leave in a vibration-free part of the laboratory for 2–3 hours to allow cells to pan gently at room temperature.

Protocol 6. *Continued*

4. Wash panning plate very gently using a pipette only, not a suction line. Remove cells, add 5 ml of PB to one edge of the dish held at a 30° angle, gently roll around 2–3 times and remove PB from the opposite edge of the dish.

5. Repeat washing 3–5 times.

6. Observe efficiency of washing under inverted microscope, gently roll dish on stage and check for the general number of free floating cells still remaining. Continue washing until no floaters remain.

D. *Preparation of Hirt*

1. Lyse the specifically panned cells *in situ* in 400 μl Hirt squirt (0.8% SDS/ 10 mM EDTA). Gently swirl around the dish to efficiently cover.

2. Cut 1–2 mm from the end of a Gilson P1000 tip and pipette the lysate gently into an Eppendorf tube (avoid shearing the genomic DNA).

3. Add 100 μl of 5 M NaCl, mix gently by inversion and place in a bath of wet ice for at least 1 hour to allow precipitation of high molecular mass primate genomic DNA.

4. Recover episomal DNA by spinning out the white genomic DNA precipitate in a minifuge for 5 min.

5. Remove clear supernatant to a fresh tube and respin if any part of the precipitate carries over.

6. Remove clear supernatant to a fresh tube.

7. Extract with 0.5 ml phenol, vortex for 1 min, spin (13 000 *g*, minifuge), and remove aqueous phase to a fresh tube.

8. Extract with 0.5 ml chloroform, spin (13 000 *g*, minifuge), and remove aqueous phase to a new tube.

9. Add 5 μg of LPA carrier, mix.

10. Add 2 volumes of ethanol, mix, and place on dry ice for 10 min.

11. Spin for 3 min, wash twice with 70% ethanol, remove residual ethanol with a P200 tip, and dissolve Hirt in 50 μl TE (for details of TE see *Protocol 1*).

E. *Transformation of Hirt*

- 10–30% (5–15 μl) of the recovered episomal DNA is transformed into 0.5 ml highly competent MC1061/p3 (> 10^8 colonies/μg).

- Plate on one 24 × 24 cm LB agar plate containing ampicillin and tetracycline, each at 10 μg/ml.

- A yield of 10^3–10^4 bacterial colonies should be obtained for the first round Hirt.

Protocol 7. Screening cDNA libraries by transient expression, panning, and episomal rescue—*Round 2*

This protocol describes protoplast fusion, expression, screening, panning, episomal rescue, and transformation.

A. *Expansion and amplification of Round 1 population (from* Protocol 6*)*

1. Scrape the bacterial population from Round 1 into a slurry in 20–50 ml LB containing ampicillin and tetracycline, each at 10 μg/ml. Titre a 1/10 or 1/100 dilution at OD_{600}.

2. Grow this population in 100–200 ml liquid culture with vigorous shaking at 37°C from a starting inoculum of $OD_{600} = 0.1$ to $OD_{600} = 0.5$.

3. Amplify plasmids by overnight incubation, with shaking, in the presence 100 μg/ml spectinomycin.

4. Prepare COS cells now for protoplast fusions the next day. Trypsinize COS cells and plate at 50–75% confluency in 10 cm culture dishes; 2 × 10 cm dishes per 100 ml of bacterial culture.

B. *Conversion to protoplasts*

The overnight liquid culture is converted to protoplasts by sequential osmotic shock, EDTA chelation, and lysozymal digestion.

1. Bacteria are pelleted (Beckman JA14/GSA rotor, 250 ml bottles) for 5 min at 10 000 r.p.m. (10 000 *g*).

2. Resuspend the bacterial pellet in 5 ml cold 20% sucrose, 50 mM Tris, pH 8.0.

3. Add 1 ml of 10 μg/ml lysozyme freshly dissolved in 250 mM Tris, pH 8.0.

4. Incubate at 4°C for 5 min.

5. Add 2 ml cold 0.25 M EDTA, pH 8.0.

6. Incubate at 4°C for 5 min.

7. Add 2 ml of 50 mM Tris, pH 8.0.

8. Incubate at 4°C for 5 min.

9. Place in a 37°C waterbath for 5 min.

10. Place on ice.

11. Check for per cent conversion to spheroplasts by microscopy (should be 90% conversion of rod-shaped bacteria to spherical protoplasts).

C. *Protoplast fusion*

Perform all manipulations in a laminar-flow hood.

1. Add 20 ml cold DME containing 10% sucrose and 10 mM $MgSO_4$ slowly, dropwise from a 25 ml pipette, swirling all the time.

Protocol 7. *Continued*

2. Remove media from 10 cm dishes of COS cells at 50–75% confluency.

3. Add 15 ml of the spheroplast slurry to each dish.

4. Place dishes in the bottom of buckets of a benchtop centrifuge (Beckman GPR, Sorvall RC6000) with rubber bases still in. Two dishes/bucket can be accommodated.

5. Spin at 2500 r.p.m. (1300 g) for 10 min at 4°C and decelerate without brake to avoid disruption of protoplast skins.

6. Aspirate fluid from dishes.

7. Add 5 ml of 50% (w/w) polyethylene glycol (PEG) 1000 or PEG 1450/50% DME into the centre of the dish.

8. After the PEG has been added to the last dish, prop all the dishes up on their lids so that the PEG drains to the bottom edge.

9. Aspirate off.

10. Leave for fusion to occur over 90–120 sec (PEG1000) or 120–150 sec (PEG1450).

11. Stop fusion by adding 5 ml of DME into the centre of the dish. The PEG layer will be swept radially away by the medium.

12. Aspirate and repeat the washing.

13. Aspirate and add 10 ml of DME/10% calf serum containing 10 µg/ml gentamicin sulfate, and leave for 4 hours.

14. Swirl the dishes to disrupt the protoplast skins, aspirate, and change the medium. Gentamicin sulfate is essential for these cultures as the residual bacterial population is so massive, penicillin and streptomycin are completely ineffective.

D. *Expression, screening, and panning*

1. Leave for 36–48 hours to allow transient expression of plasmid encoded products.

2. Repeat Mab screening and panning Round 1 (*Protocol 6*).

3. Prepare Hirt DNA, extract, precipitate, and transform in MC1061/p3. The yield should be 10^3–10^4 colonies.

Protocol 8. Screening cDNA libraries by transient expression, panning, and episomal rescue—*Round 3*

This protocol describes a further round of protoplast fusion, the subsequent transfection of the plasmid DNA into COS cells, and staining and scoring methods.

1. Perform a further round of protoplast fusion as above (*Protocol 7A–D*).

2. At the end of this round, transform 10% of the final Hirt DNA into 50 µl of competent cells, and plate on to a 10 cm Petri dish of LB containing ampicillin and tetracycline (*Protocol 7A*, step 1).

3. Incubate overnight.

4. Pick off 10–30 individual bacterial colonies into 2.5 ml of LB (+amp/tet), and grow at 37°C with vigorous shaking to saturation (8 hours minimum).

5. Isolate plasmid DNA by standard alkaline/SDS lysis methods.

6. Transfect 10–30% of the miniprep DNA into COS cells in 6-well plates by DEAE–dextran facilitation protocol (*Protocol 6A*).

7. Screen the COS cells *in situ*, 48 hours later, with the Mab at 4°C for 30 min, wash three times, and stain with a 1:100 dilution of goat-anti-mouse FITC second antibody, wash three times, and fix with PBS/2% formaldehyde.

8. Score individual wells by fluorescence microscopy. The positive clones can vary from 10–100% depending on many variables, including how abundant the original cDNA was in the library, the affinity of the antibody or ligand, and the overall efficiency of the three rounds of expression, panning, and rescue.

---

3. Functional assays of cDNA transfectants

As described in Section 1.2, one of the advantages of the transient cloning system is that functional experiments on the cloned cDNA molecules can begin immediately. Transient expression of pure cDNA clones in COS cells can lead to the accumulation of up to 10^6 molecules per COS cell surface, so that adhesion assays can be performed directly on the cells (*Figure 3*).

Protocol 9. Adhesion assays on cloned cDNAs transiently expressed in COS cells

1. Trypsinize COS cell stocks and replate at a density of $1 \times 10^4/cm^2$.

2. Transfect 10–20 µg plasmid DNA into the COS cells 4 hours at 37°C using the DEAE–dextran method (*Protocol 6A*).

3. Leave cells for a further 18 hours, trypsinize, and replate at a density of $10^4/cm^2$ on chosen assay format; 6-well, 24-well, or 96-well plates, or 3 cm or 6 cm dishes. It is essential to ensure that the cell density is correct and that the distribution is even throughout the well or dish. To avoid cells 'piling-up' in the centre of the plate, gently rock the dishes every 1–2 hours for 6 hours after replating to redistribute the cells.

Protocol 9. *Continued*

4. Measure transient expression of the encoded cDNA, this can be performed 48 hours post-transfection, but is optimal if left for 72 hours. Cell surface molecule expression can be monitored by immunocytochemistry using specific monoclonal antibodies, or by functional adhesion assay.

5. Perform functional assay, this can be achieved using radioisotopically labelled cells (overnight incorporation of [³H]thymidine) or unlabelled cells coupled with a visual assessment of adhesion photomicroscopically after fixing and staining COS cell/test cell rosettes in 0.2% crystal violet in 10% phosphate-buffered formalin (pH 7.4).

6. Allow test cells to adhere to COS transients for up to 1 hour.

7. Wash 3–5 times, monitoring for removal of floating cells.

8. Fix in PHBS/2% formaldehyde.

NB Fixed cell rosettes can be directly visualized and photographed under phase contrast using an inverted microscope.

---

4. Use of soluble recombinant adhesins

4.1 Basic outline of soluble recombinant adhesin strategy

An additional strategy for studying cell surface molecule interactions is to construct soluble forms of their extracellular domains (ECD) either alone, by introducing stop codons at the ECD/transmembrane junction, or making fusion proteins. We have had much success with the latter strategy by constructing ECD-chimeras with the Fc region of human immunoglobulin IgG1 (hinge–CH2–CH3). The advantages of this approach are: first, consistently high yields of recombinant ECD–Fc fusion proteins (up to 1 mg of pure protein from 100 ml of culture supernatant in 3 days) and a convenient, generic purification strategy by affinity isolation on Protein A Sepharose. The normal function of immunoglobulin is to be secreted at high levels from the mature plasma cell, and the Fc region appears to be well suited as a 'surrogate mother', accepting domains from other proteins and efficiently directing them through the endoplasmic reticulum and secretory pathway. Other groups have observed that Fc fusion proteins give 10–50-fold better yields of secreted proteins compared with an equivalent ECD-stop protein only (e.g. ICAM-1 stop compared with ICAM1-Fc). The Fc region allows a single purification column, avoiding the need to make ECD-specific antibody affinity columns.

4.2 Details of soluble adhesin production strategy

A vector has been constructed, designated pIG-1 (*Figure 4*), based on pCDM8, containing the genomic Fc region of human IgG1 (hinge–CH2–

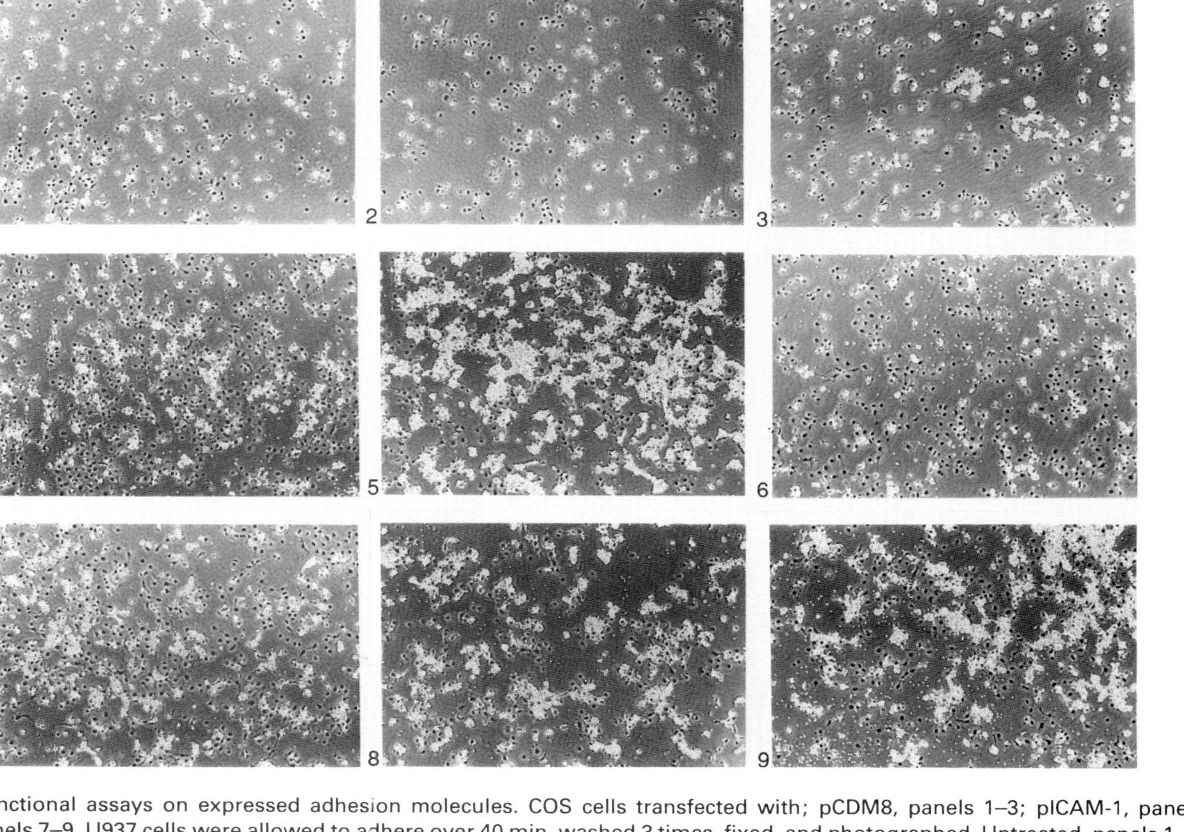

Figure 3. Functional assays on expressed adhesion molecules. COS cells transfected with; pCDM8, panels 1–3; pICAM-1, panels 4–6; pICAM-3, panels 7–9. U937 cells were allowed to adhere over 40 min, washed 3 times, fixed, and photographed. Untreated, panels 1, 4, 6; 20 μM ATP added to U937 cells 5 min before end of assay, panels 2, 5, 8; COS cells pretreated with 20 μg/ml anti-ICAM-1 blocking antibody U937 rosettes can be seen on pICAM-1 and pICAM-3 expressing COS cells.

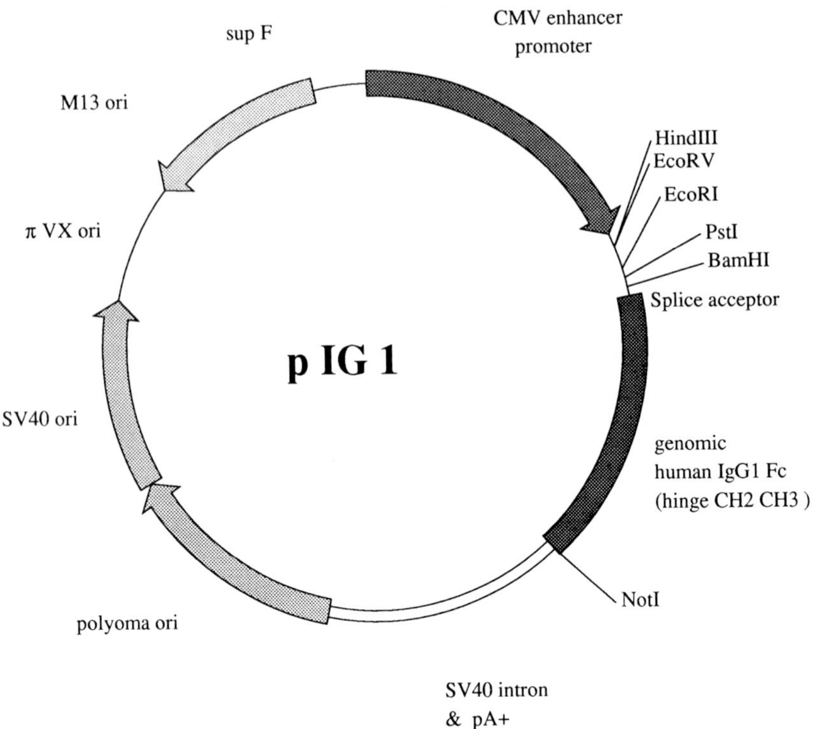

Figure 4. pIG-1. Vector for production of soluble, secreted Fc chimeric proteins.

CH3). ECD regions, either as whole ECDs or single domains are PCR amplified with appropriate restriction sites allowing directional cloning into the polylinker/stuffer region in pIG-1. A splice donor site is incorporated at the 3′ end of the PCR product so that the resulting pre-mRNA is spliced to yield a mature mRNA with the ECD directly abutting the hinge region of the IgG1 Fc. On translation, this will produce a fused chimeric protein with the ECD N-terminal to the hinge–constant 2–constant 3 region of IgG Fc.

Protocol 10. Production of recombinant protein

1. Transfect 10–20, 15 cm dishes of COS cells at 50–75% confluency with 10 µg plasmid DNA per dish by the DEAE–dextran method (*Protocol 6A*).

2. Change medium 24 hours post-transfection, replace with 25 ml DME containing 1% fetal calf serum per dish.

3. Leave for 3–6 days to allow sr-adhesins to accumulate.

4. Spin out cell debris (Beckman JA14/GSA rotor, 250 ml bottles) 5 min 10 000 r.p.m. (10 000 g), filter through a 0.45 µm Falcon bottle-top filter.

5. Add 2 ml 50% slurry of Protein A Sepharose (Pharmacia) and agitate or stir overnight.

6. Filter out supernatant through a 0.45 μm bottle-top filter, wash beads with PBS, and remove beads to a disposable 10 ml plastic column.

7. Elute sr-adhesins with 2 ml of pH 4.0 elution buffer, collecting eluate immediately into 10% volume of 1 M Tris-base, to neutralize.

8. Concentrate and buffer exchange by centrifugal membrane dialysis.

9. Label protein by incubating transfected cells with 10 μCi/ml of [^{35}S]methionine/cysteine mix in methionine/cysteine free-media (*Figure 5*).

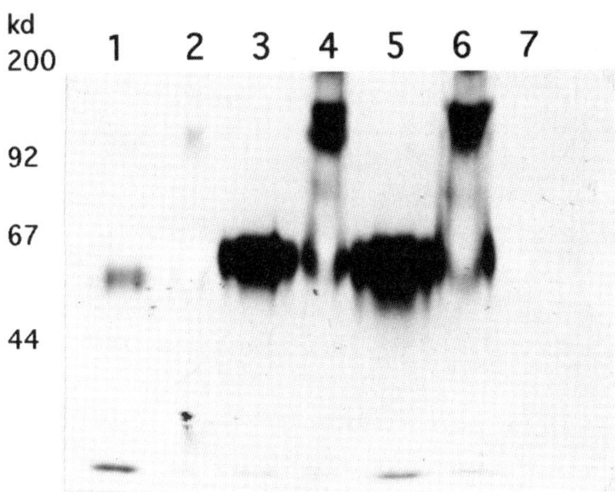

Figure 5. Example of purified soluble recombinant adhesins. ^{35}S-labelled proteins analysed on 10% SDS–PAGE. Lanes 1, 2, ICAM-1Fc; lanes 3, 4, ICAM-2Fc; lanes 5, 6, ICAM-3Fc; lane 7, vector only control; under reducing (lanes 1, 3, 5, 7) or nonreducing (lanes 2, 4, 6) conditions. Each of these ICAM-Fc chimeras consisted of the N-terminal 2 Ig-related domains only, fused to IgG-1Fc.

4.3 Examples of functional assays of homotypic and heterotypic adhesins

The sr-adhesins can be used in simple rapid solid-phase adhesion assays to assess the relative binding affinities and localize binding domains. The potential of this approach is best illustrated by specific examples; one homotypic and one heterotypic.

Protocol 11. Solid-phase adhesion assays

1. Treat 25-well assay dishes (Sterilin) with 1 ml/well of 10% toluene in ethanol for 30 min, wash with ethanol then with PBS. This removes oily plasticizers incorporated into the plastic during manufacture.

Protocol 11. *Continued*

2. Incubate with 0.5 ml/well of sr-adhesins at 50 μg/ml in 50 mM Tris, pH 9.5, for 2–4 hours.

3. Remove and wash 3 times in PBS.

4. Block remaining sites with PBS containing 2 mg/ml BSA overnight.

5. Label cells overnight in medium containing 0.2 μCi/ml [^{3}H]thymidine.

6. Wash cells in assay buffer (Hepes buffered RPM1, 1 mg/ml BSA).

7. Add 0.5 ml cells at 4×10^5/ml.

8. Incubate at 37°C for 30–40 min.

9. Wash wells 3–5 times in assay buffer.

10. Lyse 0.5 ml bound cells in 0.1 M NaOH/1% SDS.

11. Add 5 ml scintillant, mix and count in a scintillation counter.

4.3.1 Homotypic interactions

CD31 is an abundant cell surface molecule expressed on vascular endo-thelium, platelets, and monocytes. It is a member of the immunoglobulin superfamily and its ECD consists of six Ig-related domains. CD31-Fc was constructed as above and used to coat 25-well Sterilin assay plastic dishes at varying concentrations, 5–50 μg/ml/well. ^{3}H-Labelled COS cells transiently expressing CD31 were added to these dishes and allowed to adhere. After gentle washing, specifically bound cells were lysed in 1% SDS and quantitated by scintillation assay. Control surfaces were coated at equal density, with MUCD18-Fc and CD33-Fc, and other Ig-SF members. As can be seen (*Figure 6*) the CD31 transfectants specifically bound to the CD31 coated surfaces and not to control surfaces. COS cells transiently expressing MUC18 were also included in this assay and incubated with MUC18-Fc coated wells. No cells specifically adhered to these surfaces, demonstrating that MUC18 is not homotypic.

4.3.2 Heterotypic interactions

Intercellular adhesion molecules 1, 2, and 3 (ICAMs 1, 2, and 3) are ligands for Leukocyte Function-associated Antigen 1 (LFA-1). The ICAMs are members of the Ig-SF, whereas their counter-receptor is an integrin. The sr-adhesin strategy has been used to study the relative binding avidities of ICAMs 1, 2, and 3 for LFA-1, by coating plastic surfaces with sr-ICAM-1, -ICAM-2, and -ICAM-3, and quantitating the adhesion of [^{3}H]thymidine labelled LFA-1 + cells (U937 and THP-1, both human promonocyte cell lines). Control surfaces were coated with sr-CD14, which is not a cell–cell adhesion molecule. Specific cells bound to the ICAM surfaces (*Figure 7*), and ICAM-1 and -2 have an apparent stronger binding ability compared with

CD31 is (at least) a homotypic adhesion molecule

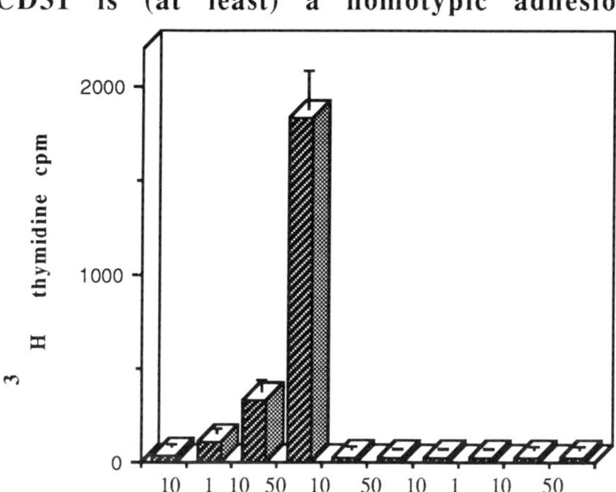

Figure 6. Use of soluble recombinant adhesins. Homotypic interactions. CD31–CD31. ^{3}H-labelled COS cells expressing CD31 adhering to CD31-Fc coated plastic surfaces, but not to CD33-Fc of MUC18-Fc coated surfaces.

ICAM-3. In addition, as the ECD-Fc constructs used here consisted only of the 2 N-terminal immunoglobulin domains of the ICAMs, this experiment localizes the LFA-1 binding site to these two domains.

5. Appendix

Details of pIG vector construction are given below.

- IgG1 gene fragment was amplified from 250 ng human genomic DNA from K562 cells.

  PCR conditions: 94 °C for 10 min, add Taq, 94 °C for 30 sec, 55 °C for 1.5 min, 72 °C for 1 min × 4: 94 °C for 30 sec, 55 °C for 30 sec, 72 °C for 1 min × 24
  Primers: GENBANK HUMIGCC4 sequence
  Forward amplification primer @ 520–547 bp
  *Bam*HI
  5'TC GGATCC GGAGGGAGGG TGTCTGCTGG AAGCAGG 3'
  Reverse amplification primer @ 2000–1970 bp
  *Not*I
  5'TGATCGCGGCCGC TGCCTCCCTC ATGCCACTCA GGCCTCAG 3'

- *pIG Construction*
  PCR products were cut with *Bam*HI/*Not*I, cloned into Bluescript (pBluescript II SK) and sequenced on the forward strand only using sequence specific primers at 200 bp intervals. No differences from the Genbank sequence were seen.

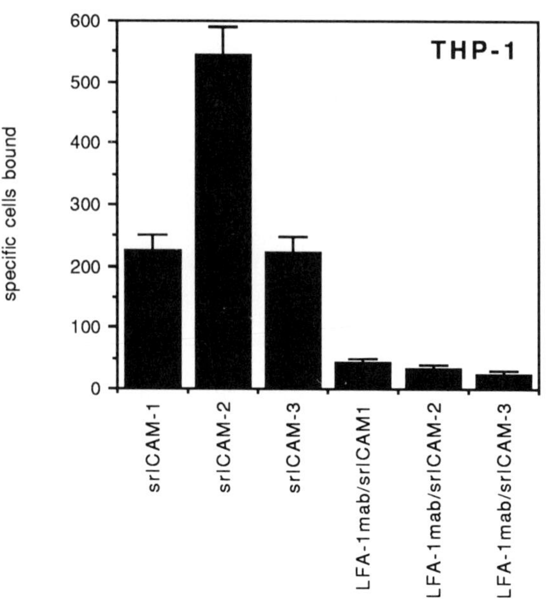

THP-1
600
500
400
300
200
100
0
specific cells bound
srICAM-1
srICAM-2
srICAM-3
LFA-1mab/srICAM1
LFA-1mab/srICAM-2
LFA-1mab/srICAM-3

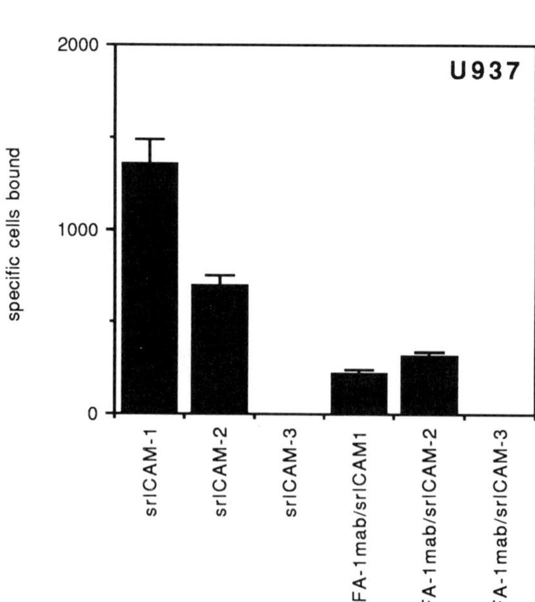

U937
2000
1000
0
specific cells bound
srICAM-1
srICAM-2
srICAM-3
LFA-1mab/srICAM1
LFA-1mab/srICAM-2
LFA-1mab/srICAM-3

Figure 7. Use of soluble recombinant adhesins. Heterotypic interactions. LFA-1/ICAM-1, 2, 3. Adhesion of LFA-1 + ^{3}H-labelled THP-1 and U937 cells to wells coated with srICAMs in solid phase. srCD14 was included as a negative control. Assays were performed in quintuplicate and results presented as means ± 1 standard deviation. Results are presented as specific cells bound, subtracted for background. The adhesion is blocked by an anti-LFA-1 monoclonal antibody.

- The single *Bam*HI site in pCDM8 was removed by mung bean nuclease digestion, and religation.
- The Fc genomic segment was moved as a *Hin*dIII/*Not*I fragment from Bluescript and ligated into *Hin*dIII–*Not*I cut *Bam*HI-minus pCDM8. This brings additional cloning sites at the 5′ ECD cloning site; *Hin*dIII–*Eco*RV–*Eco*RI–*Pst*I–*Sma*I–*Bam*HI.
  *Pst*I and *Sma*I sites cannot be used for cloning as they occur in the Fc and/or the vector.
- ECD PCR constructs are then ligated in as *Hin*dIII/*Bam*HI or *Bgl*II fragments, with splice donor sites just upstream of the 3′ enzyme site. ECD sequences will then be spliced in frame into the IgG1 hinge.
- ECD-Fc chimeras will dimerize via this hinge. Splice donor sequence at 3′ end of ECD
  PCR product
    splice donor   *Bam*HI
  5′ XXX ACA G   GT AAG TGG ATC CGT 3′
  3′ YYY TGT C   CA TTC ACC TAG GCA 5′
  (Vertebrate 5′ splice site consensus: 5′ C/AAG GTAAGT 3′).

6. Concluding remarks

Transient expression cloning allows the rapid isolation of cDNA clones encoding cell surface molecules. In addition, the high levels of expression permit immediate functional analysis, and production of soluble recombinant adhesins enables quantitation of binding affinities and dissection of binding domains.

Acknowledgements

I thank members of the Cell Adhesion Laboratory: Elizabeth Barber, Dr Stuart Bloom, Dr Jonathan Fawcett, Dr Sylvie Freeman, Dr Claire Holness, and Dr Jack McBain; Dr Lindsey Needham at British Biotechnology, Oxford for expertise in functional analysis of transfected clones. This work was supported by the Imperial Cancer Research Fund, UK.

References

1. Seed, B. and Aruffo, S. (1987). *Proc. Natl Acad. Sci. USA*, **84**, 3365–9.
2. Aruffo, S. and Seed, B. (1987). *Proc. Natl Acad. Sci. USA*, **84**, 8573–7.
3. Seed, B. (1987). *Nature*, **329**, 840–2.
4. Linsley, P., Clark, E., and Ledbetter, J. (1989). *Proc. Natl Acad. Sci. USA*, **87**, 5031–5.
5. Simmons, D., Makgoba, W., and Seed, B. (1988). *Nature*, **331**, 624–7.
6. Staunton, D., Dustin, M., and Springer, T. (1989). *Nature*, **339**, 61–4.
7. Bevilacqua, M., Stengelin, S., Gimbrone, M., and Seed, B. (1989). *Science*, **243**, 1160–5.
8. Lowe, J., Stoolman, L., Nair, R., Larsen, R., Behrend, T., and Marks, R. (1990). *Cell*, **63**, 475–84.
9. Walz, G., Aruffo, A., Kolanns, W., Bevilacqua, M., and Seed, B. (1990). *Science*, **250**, 1132–5.
10. Stamenkovic, I. and Seed, B. (1990). *Nature*, **345**, 74–7.
11. Stamenkovic, I., Amiot, M., Pesando, J., and Seed, B. (1989). *Cell*, **56**, 1057–62.
12. Stamenkovic, I., Clark, E., and Seed, B. (1989). *EMBO J.*, **8**, 1403–10.
13. Aruffo, S., Stamenkovic, I., Melnick, M., Underhill, C., and Seed, B. (1990). *Cell*, **61**, 1303–13.
14. Camerini, D., James, S., Stamenkovic, I., and Seed, B. (1989). *Nature*, **342**, 78–82.
15. Oquendo, P., Hundt, E., Lawler, J., and Seed, B. (1989). *Cell*, **58**, 95–101.
16. Berendt, A., Simmons, D., Tansey, J., Newbold, C., and Marsh, K. (1989). *Nature*, **341**, 57–9.
17. Osborn, L., Heission, C., Tizard, R., Vassallo, C., Luhowskyi, S., Chi-Ross, G., and Lobb, R. (1989). *Cell*, **59**, 1203–11.
18. Elices, M., Osborn, L., Takada, Y., Grouse, C., Luhowskyj, S., Hemler, M., and Lobb, R. (1990). *Cell*, **60**, 577–84.
19. Yamasaki, K., Taga, T., Hinta, Y., Yawata, H., Kawanishi, Y., Seed, B., Taniguchi, T., Hirano, T., and Kishimoto, T. (1988). *Science*, **241**, 567–9.
20. Gearing, D., King, J., Gough, N., and Nicola, N. (1989). *EMBO J.*, **8**, 3667–76.
21. Chirgwin, J. M., Pryzbyla, A. E., MacDonald, R. J., and Rutter, W. J. (1979). *Biochemistry*, **18**, 5294–302.
22. Gubler, U. and Hoffman, B. J. (1983). *Gene*, **25**, 263–75.
23. Kioussis, D., Wilson, F., Daniels, C., Leveton, C., Taverne, J., and Playfair, J. (1987). *EMBO J.*, **6**, 355–61.
24. Hauer, C., Getty, R., and Tykocinski, M. (1989). *Nucleic Acids Research*, **5**, 1989–2003.
25. Elliott, J. F., Albrecht, G. R., Gilladoga, A., Handunnetti, S. M., Neequaye, J., Lallinger, G., Minjas, J. N., and Howard, R. J. (1990). *Proc. Natl Acad. Sci. USA*, **87**, 6363–7.
26. Gluzman, Y. (1981). *Cell*, **23**, 175–80.
27. Gorman, C. (1986). In *DNA cloning* (ed. D. M. Glover), Vol. 11, pp. 143–90. IRL Press, Oxford.
28. Gorman, C., Rigby, P., and Lane, D. (1985). *Cell*, **42**, 519–26.
29. Aruffo, A. and Seed, B. (1987). *EMBO J.*, **6**, 3313–18.
30. Stamenkovic, I. and Seed, B. (1988). *J. Exp. Med.*, **168**, 1205–8.

31. Stamenkovic, I. and Seed, B. (1988). *J. Exp. Med.*, **167**, 1975–9.
32. Simmons, D. L., Tan, S., Tenen, D. G., Nicholson-Weller, A., and Seed, B. (1989). *Blood,* **73**, 284–9.
33. Simmons, D., Walker, C., Power, C., and Pigot, R. (1990). *J. Exp. Med.*, **171**, 2147–52.
34. Simmons, D. and Needham, L. A. (1992). In *Vascular endothelium: interactions with circulating cells* (ed. J. Gordon), pp. 3–29. Elsevier, New York.
35. Davies, A., Simmons, D., Hale, G., Harrison, R., Tighe, H., Kachmann, J., and Waldman, H. (1989). *J. Exp. Med.*, **170**, 637–54.
36. Simmons, D. L. and Seed, B. (1988). *Nature,* **333**, 568–70.
37. Stengelin, S., Stamenkovic, I., and Seed, B. (1988). *EMBO J.*, **7**, 1033–60.
38. Allen, J. and Seed, B. (1989). *Science,* **243**, 378–80.
39. Simmons, D., Satherthwaite, A. B., Tenen, D. G., and Seed, B. (1992). *J. Immunol.*, **148**, 267–71.
40. Simmons, D. and Seed, B. (1988). *J. Immunol.*, **141**, 2791–800.

6

Looking for protein–protein interactions

CAROL A. MIDGLEY and DAVID P. LANE

1. Introduction

The specific interactions of protein molecules to form homo- and hetero-oligomers underlie many key events in biology, from the assembly of viruses to the formation of multisubunit protein machines of great complexity. Traditionally the study of such interactions has demanded the availability of large amounts of purified components, and the members of a complex have been identified through their extensive copurification. These techniques are restricted to the analysis of tight associations between relatively abundant proteins, and cannot easily be used to study the biological regulation of a protein complex in different cellular environments.

More recently, powerful methods have been developed that permit the study even of weak protein–protein interactions when pure proteins are not available, where only one partner in the complex is known, and when the components are present as only trace amounts in complex mixtures. All of these new methods rely on sensitive and specific detection and isolation of at least one partner in the complex. This identification is based either on the production of specific antibodies to one component of the complex, or the isolation and specific labelling of one component (*Figure 1*). While this labelling process has traditionally consisted of the covalent modification of purified protein, recombinant DNA methods have allowed a novel approach in which genetic engineering is used to 'tag' the protein. This is totally compatible with gene expression systems and can provide a very effective way for producing easily detectable protein probes as required. The protein tag may be an 'epitope tag' that consists of a short stretch of precise amino acid sequence recognized by a specific monoclonal antibody, it may be a short stretch of amino acids that represents a ligand for specific affinity resins or a site for covalent modification, or it may be a larger protein domain that contains defined ligand-binding sites.

Certain constraints are common to all these methods. First it is important that the labelling method does not fundamentally disturb the interactions of the labelled protein with its target proteins. Second, specific complex formation is subject to the laws of mass action, and a given affinity of interaction can only be detected within a defined range of component concentrations. Finally, and most importantly, there is a great need to establish criteria for the specificity and biological relevance of complex formation. It is in this area that these new methods are subject to abuse and misinterpretation. In particular, the highly specific assays achieved for some of the labelled probes, combined with the very high concentrations of potential partners that can be obtained using recombinant DNA expression systems can give very misleading results. Most proteins have complex surfaces displaying variant local charge, and all proteins like to stick to each other to some extent. The main defence against these artefacts are careful quantitation and titration of the reactive components, and direct demonstrations of selectivity in competition with other realistic ligands (not just bovine serum albumin!). In the ultimate analysis, however attractive a potential association may seem there is no substitute for clear biological data demonstrating its *in vivo* significance. These problems of specificity and significance have become even more acute in the last couple of years with the recognition of the existence of simple structural motifs associated with protein–protein interactions, such as the leucine zipper (or coiled-coil motif), the Retinoblastoma protein pocket domain and its peptide ligand, and the SH (Src homology) domain. Variations of these simple motifs are used by many protein complexes as a kind of molecular 'lego' brick. Specificity is achieved for these motifs by their precise variation

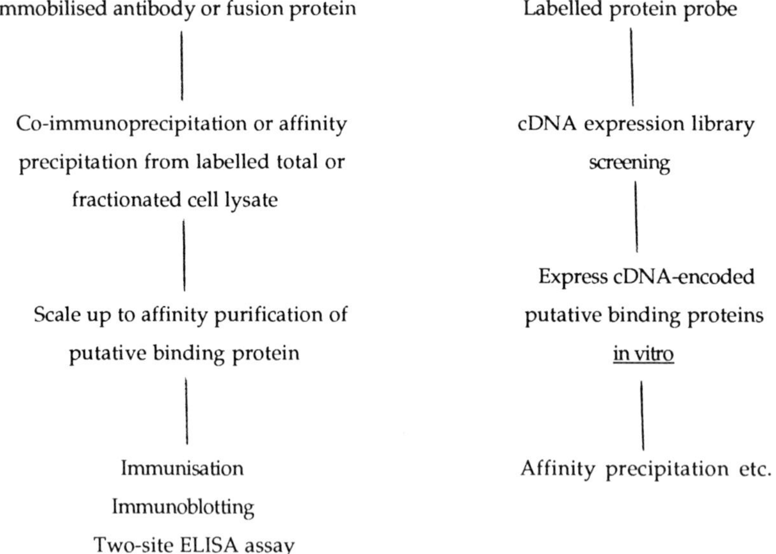

Figure 1. Flow diagrams of steps towards defining protein–protein interactions.

and context. When this is circumvented by non-competitive environments or the use of high specific activity probes and high-level expression systems, a subtle sort of nonsense emerges in which 'specific' interactions occur that probably never take place *in vivo*. What follows are a set of practical protocols detailing the methods we have found most useful and noting some pitfalls. Happy hunting!

2. Bacterial fusion proteins as affinity substrates for purifying protein complexes

Where an appropriate antibody is available to one component of a protein complex, immunoprecipitation experiments (see Section 5) will often reveal copurifying associated proteins. As an alternative to using specific antibodies to capture protein complexes, you can apply many of the same techniques directly to a purified protein 'probe' furnished with an affinity tag which allows its immobilization on solid-phase beads. The most convenient route to achieving this is to express the protein in *Escherichia coli* as a fusion to the affinity tag. Numerous plasmid vectors are now available for expressing fusion polypeptides which can be purified from lysed *E. coli* cells in one step by substrate affinity or immunoaffinity strategies (see *Table 1*). Alternatively, you can make your own 'epitope tag' by introducing, into the amino- or carboxy-terminal region of your gene sequence, synthetic linkers which encode a defined antibody epitope as described for the pFlag vector (see *Table 1*) or the 'Myc tag' recognized by an anti-Myc antibody called 9E10 (6). An epitope tag provides a means of both affinity purifying and detecting the protein using an antibody.

This section will describe the use of just one of these techniques, the glutathione S-transferase (GST) gene fusion system. The plasmids pGEX-2T

Table 1. Plasmids for expression of fusion proteins with affinity tags

Plasmid	Affinity tag fusion partner	Affinity resin	Reference
pRIT2T (Pharmacia)	staphylococcal protein A	IgG sepharose	1
pMAL1 (Pharmacia)	maltose-binding protein	amylose resin	2
pGEX-2T or -3X (Pharmacia)	glutathione-S transferase	glutathione sepharose	3
pQE (Quiagen, Inc.)	6 × HIS tag	nitriloacetic acid resin (nickel chelator)	4
pFlag1 (International Biotechnologies Inc.)	'Flag' antibody epitope	anti-Flag sepharose	5

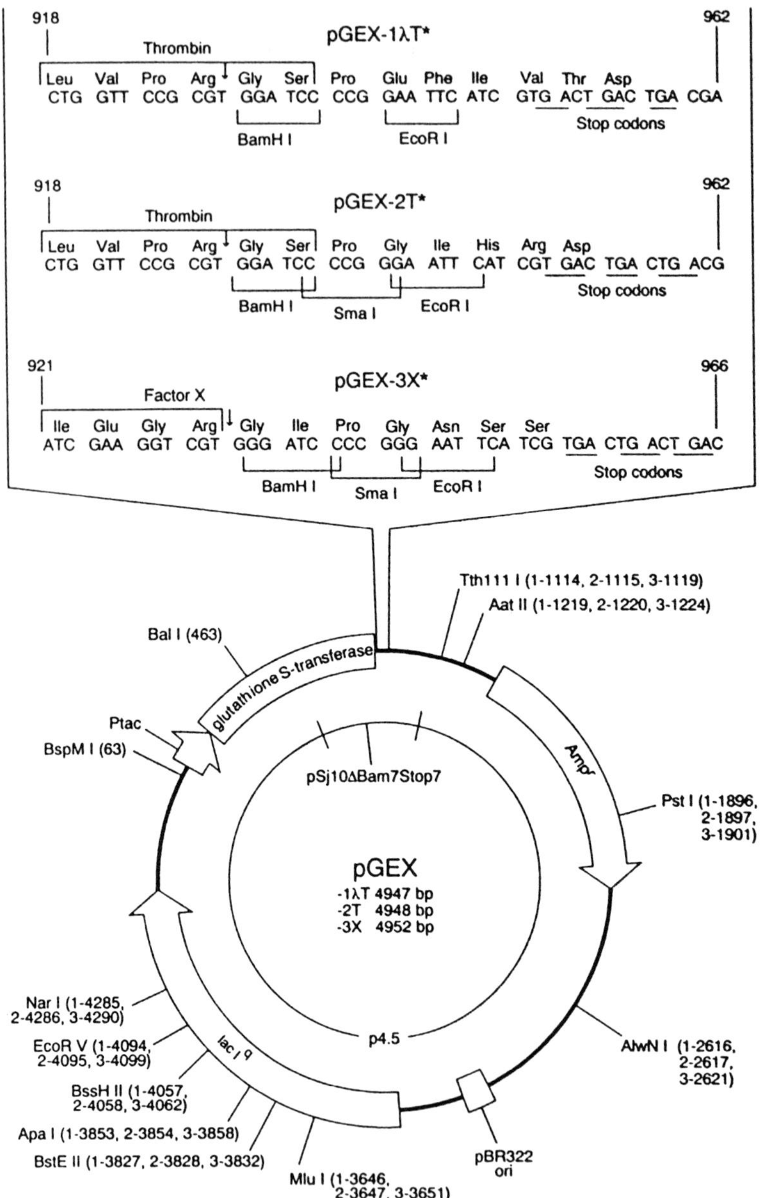

*For restriction sites listed in the body of the plasmid map, the designations of 1, 2, 3 preceding the restriction site location refer to pGEX-1λT, pGEX-2T and pGEX-3X, respectively.

Figure 2. Plasmid map of pGEX fusion vector series. (Reproduced with the kind permission of Pharmacia P-L Biochemicals Inc.)

and pGEX-3X (*Figure 2*) are induced by the addition of isopropyl β-D-thiogalactoside (IPTG) to produce high levels of genes or gene fragments as fusions with the *Schistosoma japonicum* glutathione S-transferase (3). The fusion proteins are purified from lysed cells by affinity chromatography using glutathione Sepharose (Pharmacia or Sigma). The protein-coated beads can then be used in affinity precipitation with cell lysates (7) or for affinity purification. An added advantage of this system is the build-in procedure for the elution of GST-fusion proteins by competition with free glutathione. This versatile system can, therefore, provide either affinity-purified protein complexes or the pure uncomplexed protein alone as a probe for library screening, immunization, etc. The vectors described in *Figure 2* also encode protease cleavage sites to allow separation of the polypeptide from the fusion protein where desirable.

You should be aware, however, that whereas antibody co-immunoprecipitation can capture existing complexes from cell lysates, you are asking your protein probe to form new associations with proteins which may already be involved in very tight associations in the lysate. A particular advantage of using recombinant protein probes, however, are the possibilities for introducing internal controls in the form of genetically altered forms of the protein. Kaelin *et al.* (7) were able to use GST-RB (retinoblastoma gene product) protein fusions with single point mutations or deletions to demonstrate that the binding of certain cellular proteins was specific to the wild-type form of RB protein.

Protocol 1. Expression of GST-fusion proteins and binding to glutathione beads

1. Set up a culture of the bacterial clone containing the recombinant plasmid in Luria broth with 50 μg/ml ampicillin. Incubate overnight at 37°C, with shaking.

2. The next day dilute the culture one in ten and grow, shaking at 37°C for another hour before adding IPTG to a final concentration of 1 mM.

3. Grow for another 3–4 h.

4. Retain aliquots of the culture with and without IPTG induction for polyacrylamide gel electrophoresis (PAGE). If fusion protein yield is low due to degradation, try delaying addition of the IPTG until less than an hour before harvesting the cells.

5. Spin down 10 ml of cells at 3000 *g* for 5 min at 4°C.

6. Remove the supernatant and resuspend the pellet in 1 ml of cold lysis buffer. You will need to determine the best conditions for lysis, but good buffers to start with include Nonidet P40 (NP-40) lysis buffer (*Table 2*) or phosphate-buffered saline (PBS) with 1% Triton X-100.

Protocol 1. *Continued*

7. Sonicate the cells with brief bursts to lyse them. Keep the lysate cool on ice and use minimum sonication to reduce damage to proteins and minimize background problems.

8. Spin the lysate at 10 000 *g* for 5 min at 4°C.

9. Transfer the supernatant to a clean tube and add 20–50 µl of a 50% v/v suspension of glutathione beads which have been washed several times and resuspended in lysis buffer containing 0.5% dried milk protein as a blocking agent.

10. Rock for 15 min at room temperature.

11. Wash three times with 1 ml of cold lysis buffer. Pellet beads each time at 10 000 *g* for 15 sec and remove washes by aspiration using a 23-gauge needle and syringe.

12. Analyse some of the original cell culture for expression of the fusion protein and an aliquot of the beads for the efficiency of binding.

---

3. *In vitro* labelling of pure proteins

Screening proteins immobilized on nitrocellulose or nylon filters after affinity blotting (see Section 8) or as cDNA expression libraries (see Section 9) can be accomplished with labelled protein. It is beyond the scope of this chapter to detail all the available methods of labelling and detecting proteins. We have used the GST-fusion expression system to purify protein for cDNA expression library screening by labelling the fusion protein *in vitro* with [125]iodine by a procedure (see *Protocol 2*) which minimizes the chances of oxidative damage to the protein (8), but it may be necessary to assess different techniques, especially where the retention of protein-binding activity can be tested on a known interaction after the labelling step. Macgregor *et al.* (9), before using biotinylated Jun protein for screening an expression library, tested the ability of their labelled protein to dimerize and to detect a recombinant lambda phage expressing c-Fos.

You might consider introducing a phosphorylation site into a recombinant protein probe using synthetic oligonucleotides. This allows *in vitro* labelling to high specific activity. Blanar and Rutter (10) used c-Fos probes with an amino-terminal extension which included an in-frame five amino acid (ARG ARG ALA SER VAL) recognition element for the catalytic domain of cAMP-dependent heart muscle kinase (11). A GST-fusion plasmid pGEX-2TK incorporating this same recognition sequence has recently been described (12).

Nonradioactive direct protein-labelling techniques are widely available, e.g. biotin or digoxigenin (Boehringer) which should be used as described in the manufacturer's instructions. These bulky adducts may, however, sterically compromise complex formation, and it may be necessary to titrate the labelling reaction to achieve an acceptable balance between incorporation and retention of binding activity. Alternatively, bound protein can be detected using a labelled specific antibody (13). Blanar and Rutter (10) inserted into their fusion protein an epitope tag recognized by a commercially available antibody (anti-FLAG, IBI see *Table 1*) which can be used either for detection or for purification.

Protocol 2. Iodination with an iodogen-coated tube

1. Prepare the iodogen-coated tube; dissolve the iodogen (1,3,4,6-tetrachloro-3α,6α-diphenylglycoluril, Pierce Chemical Co.) at 0.5 μg/ml in chloroform. Dispense 20 μl aliquots into 1.5 ml glass tubes and allow to evaporate in a fume hood overnight. The tubes can be stored in a desiccator at room temperature for several years. The use of immobilized reagent minimizes damge to the protein from direct contact with the oxidant, and makes it unnecessary to add reducing agent to stop the reaction.

2. Prepare a disposable column of Sephadex G-50 medium (Pharmacia), bed volume about 6–10 ml in a syringe or pipette plugged with glass wool. Equilibrate with several bed volumes of PBS with 1 mg/ml bovine serum albumin (BSA) to block any protein-binding sites.

3. Add 20 μl of 0.1 M sodium phosphate, pH 7.2, containing 10 μg of the protein to the iodogen tube. There should be no reducing agents in this solution so dialyse your protein against sodium phosphate before use, especially if they are GST-fusions eluted with free glutathione. Add 1 mCi Na[^{125}I] (14–16 mCi/ml, ICN Flow).

4. After 10 min, quench and stop the reaction by adding 10 μl saturated L-tyrosine, 10 μl 0.5 M potassium iodide, and 50 μl PBS containing 2 mg/ml BSA.

5. Apply the protein to the Sephadex column. Elute with PBS containing 1 mg/ml BSA, collecting 0.5 ml fractions. Count the fractions and pool those containing labelled protein. Expect specific activities of 4–7 $\times$ 10^7 c.p.m./μg.

---

4. Preparation of cell lysates

Protein–protein interactions are generally noncovalent and reversible and complexes are, therefore, in equilibrium with the free components. The concentration of protein components, temperature, pH, and solvents will all affect the equilibrium constant, i.e. the number of complexes found at equilibrium, by driving the reaction toward completion or away from it. The first consideration is to keep the concentration of components as high as possible since dilution increases dissociation of weakly-associated protein complexes, so extracts should be kept to as small a volume as possible. For example, quantitative immunoprecipitations usually need antibody affinities of $> 10^8$ mol/l or higher. It may, in some cases, be preferable to carry out subcellular fractionation in order to increase the relative concentration of putative target proteins in your extract. Fractions enriched for mitochondria, nuclei, cytosol, or membranes can be prepared by centrifugation (14). Similarly, you may want to fractionate the lysate by density-gradient centrifugation. If proteins copurify in this procedure it may increase your confidence if they are involved in complexes, and may improve detection by reducing background. Labelled extracts ($\sim 10^6$ cells) in about 1 ml of lysis buffer (without sodium dodecyl sulphate, SDS) can be layered on to 5–10 ml linear 5–20% sucrose gradients (made in PBS or 10 mM Tris–HCl pH 7.3, 0.1 M NaCl, 0.5% NP-40). Gradients should be spun at 55 000 g for 16 h at 4°C. Take 0.25–0.5 ml fractions by puncturing the bottom of the centrifuge tube or pumping through a tube pushed to the bottom. Individual fractions can be subjected to immuno-precipitation and the protein complexes analysed by gel electrophoresis.

Cells grown in tissue culture are usually labelled *in vivo* with medium containing radiolabelled amino acids (Section 4.1) or [^{32}P]orthophosphate and lysed with detergents. Alternative lysis methods which use mechanical forces, e.g. sonication, can disrupt protein complexes, while freeze–thawing may cause damage due to rapid changes in pH and salt concentration.

4.1 *In vivo* labelling of cellular proteins

To look at protein complexes released from lysed cells it is usually necessary to radioactively label total cell protein *in vivo*. For eukaryotic cells grown in tissue culture, total protein can be labelled by growing cells in the presence of appropriate radioactive amino acids, usually [^{35}S]methionine or -cysteine. Incorporation is linear up to about 5 mCi/10^6 cells, therefore the number of counts incorporated depends primarily on the amount of label added. Procedures for metabolic labelling of yeast and *E. coli* proteins are not described here but can be found in reference 15.

Protocol 3. Labelling cell monolayers

1. Obtain 75% confluent cells. Always prewarm the medium to appropriate cell growth temperature before use. Remove growth medium and wash the cells once in medium without methionine (or without cysteine where appropriate).

2. Remove the wash medium and add the minimum volume of medium (without methionine) to cover the cells (2 ml/100 mm dish, 0.5 ml/60 mm dish, 0.25 ml/35 mm dish). Larger volumes may be needed for very long labelling periods.

3. Add [^{35}S]methionine. As a general guide for a 3 h label: for a protein of major abundance ($> 10^7$/cell) add 5 μCi, intermediate abundance (10^5–10^7/cell) add 50 μCi, low abundance (10^3–10^5/cell) add 500 μCi. If fluorography is used to enhance detection after gel electrophoresis, less label may be required.

4. Incubate at 37°C, or the appropriate temperature, for the desired length of time, usually 2–4 h. Incorporation will be linear for 6–8 h. For longer labelling periods medium should be supplemented with additional non-radioactive amino acids.

5. Remove medium and wash monolayer with cold PBS. Go on to cell lysis procedure.

Protocol 4. Labelling cells in suspension

1. Pellet cells at 400 g for 5 min, remove medium, and wash cells once by resuspending in prewarmed medium without methionine, and repellet.

2. Remove wash medium and resuspend cells at 10^7/ml in medium without methionine, and transfer to a small tissue culture plate.

3. Follow *Protocol 3*, steps 3 and 4.

4. Transfer cells to a centrifuge tube and pellet at 400 g for 5 min. Remove medium and wash cells once with cold PBS, repellet, and go on to the cell lysis procedure.

4.2 Detergent lysis of cells

Often the choice of lysis buffer and detergent will determine which antigens are released efficiently and whether multimeric complexes form or remain intact. Several conditions should be tested to determine which allows complex formation, particularly where interactions are weak, and also to minimize nonspecific interactions with other components of the lysate. It may be possible to choose conditions where such contaminants are not released into the extract as efficiently as are your target proteins. Variables include the choice of detergents (ionic, e.g. SDS, deoxycholic acid (DOC); nonionic, e.g. Tween; or zwitterionic, e.g. 3-[(3-cholamidopropyl)-dimethylammonio]-1-propane-sulfonate (CHAPS)), salt concentration, presence of divalent cations or metal ions, and pH. Chaotropic agents, such as urea, can only be included where they will not affect protein interactions. It is important to keep lysates cold to help reduce degradation by proteases released during extraction. Additionally you should consider supplementing your lysis buffer with protease inhibitors, these commonly include phenylmethylsulfonyl fluoride (PMSF), aprotinin, pepstatin, and ethylenediaminetetraacetate (EDTA). See *Table 2* for some typical lysis buffer recipes.

Table 2. Cell lysis buffers

NP-40 lysis buffer: 150 mM NaCl, 1.0% NP-40, 50 mM Tris–HCl (pH 8.0)

High-salt buffer: 500 mM NaCl, 1.0% NP-40, 50 mM Tris–HCl (pH 8.0)

Low-salt buffer: 1.0% NP-40, 50 mM Tris–HCl (pH 8.0)

RIPA buffer: 150 mM NaCl, 1.0% NP-40, 0.5% DOC, 0.1% SDS, 50 mM Tris–HCl (pH 8.0)

Protocol 5. Lysis of cell monolayers

1. Place the dish of washed cells on ice, after labelling, as described in *Protocol 3*. Add 1 ml of prechilled lysis buffer per 100 mm dish.

2. Incubate the plate on ice for 30 min, rocking occasionally to spread the buffer.

3. Tilt the plate and transfer buffer to a 1.5 ml centrifuge tube (cell debris can be scraped off and transferred as well if desired).

4. Spin at 10 000 *g* for 5 min at 4°C and transfer soluble supernatant to a fresh tube. Keep the lysate chilled throughout the procedure in case the proteins are sensitive to proteases.

Protocol 6. Lysis of suspension cells

1. Place tube of washed cells on ice and resuspend pellet in 1 ml of prechilled lysis buffer, after labelling, as described in *Protocol 4*.

2. Incubate on ice for 30 min, mixing occasionally.

3. Spin at 10 000 *g* for 5 min at 4°C and transfer soluble supernatant to a fresh tube.

5. Immunoprecipitation of protein complexes

Where an appropriate antibody is available specific antigens and their associated ligands, e.g. proteins or nucleic acids, can be co-immunoprecipitated from labelled cell lysates (16). Antibodies are simply added to the lysate to allow attachment to protein complexes, and then a solid-phase matrix of Sepharose beads coated with protein A or protein G is added to adsorb the immune complexes. Protein A has good affinity for rabbit polyclonal antibodies and for the commonly used mouse monoclonal IgG subclasses except IgG1. A detailed list of affinities is provided in reference 15. In problem cases you should use protein A with a second bridging layer of antibody, or replace protein A with protein G. Unbound proteins, etc. are removed by washing the beads, and the remaining bound complexes can be analysed, usually in the first instance by gel electrophoresis. Where high backgrounds are encountered try more stringent washes, e.g. high salt, 0.2% SDS, 1% Tween-20. It is most important that appropriate control antibodies are included, especially negative controls consisting of irrelevant antibodies from the same animal and preferably the same class and subtype.

This procedure can be scaled up to perform immunopurification by first chemically crosslinking the antibody to protein A beads which are then used as a column, or batch-wise for adsorbing the protein complexes from large quantities of lysate. Subsequent elution of complexes can provide material for biochemical assays or immunization.

Protocol 7. Immunoprecipitation from cell lysates

Reagents

- Laemmli sample buffer: 2% SDS, 10% glycerol, 100 mM dithiothreitol (DTT), 60 mM Tris–HCl (pH 6.8), 0.001% bromophenol blue

Method

1. 'Preclear' the cell lysate to remove proteins which bind nonspecifically to the immune complexes or beads. Lysates can be incubated with an aliquot of beads before starting the immunoprecipitation procedure, or if high backgrounds are encountered it may be necessary to preincubate with 50 μl of normal rabbit serum per ml of lysate. The rabbit antibody complexes can be removed by binding to fixed *Staphylococcus aureus* Cowan 1 (available commercially) or protein A beads. Make sure lysates have been thoroughly cleared of any insoluble material by centrifuging at 10 000 *g* before use.

2. Use either polyclonal or suitably high-affinity monoclonal antibodies for immunoprecipitation. Usually 0.5–5 μl of polyclonal serum, 10–100 μl hybridoma supernatant, or 0.1–1 μl of ascites is added to about 1 ml of lysate.

3. Incubate the lysate on ice for about an hour. Spin the lysate again to remove insoluble material before adding the beads.

4. Add (if necessary) 0.1 μg of an intermediate antibody if monoclonals are used, e.g. rabbit anti-mouse immunoglobulin, since not all Ig subclasses bind well to protein A.

5. Add 10–50 μl of protein A or G beads, this is usually sufficient to precipitate immune complexes in small volumes. It is often worth adding 50–100 μl of Sepharose CL6B beads (Pharmacia) where small quantities

Protocol 7. *Continued*

of protein A or G beads are used, to help minimize loss of beads during aspiration of wash buffer. The Sepharose and protein A or G beads should be thoroughly washed in the appropriate buffer, usually the cell lysis buffer, before use.

6. Incubate at 4°C on a mixing wheel or rocker for about 30 min. Pellet beads at 10 000 g for 15 s at 4°C. Wash the bead-bound complexes three times with lysis buffer, repelleting as described and remove washes by aspiration; a 23-gauge needle and 2 ml syringe are convenient for this.

7. Remove the final wash and assay the bound complexes. For SDS–PAGE analysis add 50 μl of Laemmli sample buffer. Heat to 85–100°C for 10 min. Spin to pellet beads, transfer supernatant to a fresh tube, and load on the gel.

---

6. Immunopurification of protein complexes

To scale up from the last procedure, antibodies can be bound directly to protein A or G Sepharose beads. Once maximum binding is achieved the antibody is covalently crosslinked to the matrix via a bifunctional crosslinking reagent, usually dimethylpimelimidate. It is essential that the optimum conditions of antibody binding and coupling to the protein A or G beads are determined by small-scale experiments. If antibodies have a particularly low affinity for protein A or G this may be circumvented by binding to an intermediate layer of anti-immunoglobulin molecules with good affinity. Alternatively, antibodies can be covalently bound directly to chemically-activated beads or the antibodies themselves can be botinylated and coupled to streptavidin–agarose beads (15).

Protocol 8. Protein A bead–antibody affinity columns

1. Bind antibody. The source of antibody can be serum, tissue culture supernatant (in medium without added serum), ascites, or purified immunoglobulin. Usually you will require about 2 mg/ml of the protein A or G beads. The most efficient technique is to gently mix the antibodies and beads at room temperature for an hour. For large volumes of tissue culture supernatant or ascites fluid you may filter through an 8 μm Millipore filter then circulate the medium through a column of beads at a flow rate of < 10 ml/h. If the antibody does not bind efficiently, attach an anti-immunoglobulin antibody to the protein A beads using a saturating concentration (20 mg/ml) before binding your own antibody.

2. Wash the beads or column with 10–20 volumes of PBS followed by two washes of 10 volumes of 0.1 M sodium borate, pH 9.0 (to wash a bead

slurry, pellet by centrifugation at 3000 *g* for 5 min or 10 000 *g* for 30 sec and remove wash by aspiration). Retain a 10 µl sample of the beads.

3. Remove all buffer and make up the coupling reagent; dissolve 0.5 *g* dimethylpimelimidate (Sigma) in enough 0.1 M sodium borate, which has not been pH adjusted, to give a pH of 9.0 (you should need about 40 ml). Add sodium borate, pH 9.0, to give a final volume of 50 ml and check the pH (coupling must be performed above pH 8.3).

4. Either resuspend the bead slurry in 10 volumes of coupling buffer and mix for 30 min at room temperature on a rocker or shaker, or recirculate 20–30 ml of buffer over the column overnight. Retain a 10 µl sample.

5. Wash the beads or column in 0.02 M ethanolamine, pH 9.0, for 10–20 min.

6. It is often a good idea to wash the beads in 0.1 M citrate after coupling to remove any unbound antibody. Finally, wash the beads and store in PBS with 0.01% merthiolate.

7. Check the efficiency of coupling by running the samples of beads taken before and after coupling on a 10% SDS–polyacrylamide gel and stain with Coomassie blue. You should see a heavy chain band (55 000 mol.wt) in the before coupling but not the after coupling sample. You are now ready to bind the antigen complex to the antibody beads.

Protocol 9. Binding antigen complexes to the antibody–beads

As when binding antibody to the beads, you may either mix the antigen with the bead slurry or pass it slowly down a column of beads keeping them in contact as long as possible. The binding buffer used can be cell lysis buffer and should be determined empirically on small batches of antibody–beads.

1. Mix the antigen solution with the antibody–beads and incubate at 4°C with constant gentle agitation (preferably rocking not shaking) for 1 h to overnight.

2. Allow the beads to settle, remove the binding buffer by aspiration, resuspend in fresh cold binding buffer, allow to settle again, and remove the wash. This helps to remove any unbound protein aggregates.

3. Transfer to a small column and wash with 20 volumes of binding buffer. Go on to elution of bound complexes.

Alternative protocol

1. Transfer the antibody beads to a small column and wash with 20 volumes of PBS.

Protocol 9. *Continued*

2. Pass the antigen solution through a column of agarose beads equal in volume to the antibody column. This will reduce nonspecific binding.

3. Apply the antigen solution to the column either by gravity flow or at a flow rate of < 2 ml/h using a pump.

4. Wash with 20 volumes of buffer and go on to the elution step.

6.1 Elution of protein complexes

You will only be able to find good conditions for the elution of antigens and antigen complexes by testing several buffers. It may be possible to use low pH (3–1.5) or high pH (10–12.5), 3–5 M $MgCl_2$, 25–50% ethylene glycol, etc. (15), but avoid reducing and denaturing conditions if you wish the protein complex to remain intact.

An alternative is to compete for antibody binding with a small molecule, e.g. a peptide which mimics the antibodies epitope. This approach has proved successful in the affinity purification of p53 protein and complexes of p53 with HSP70 and other associated proteins from rat cells using monoclonal antibody PAb421 (17). This approach is worth taking where the antibody was raised against a synthetic peptide. Often using a large excess of the peptide will achieve elution, but efficiency depends largely on the relative affinity of the antibody for the peptide versus the antigen, and also on the rate of dissociation. Often with high-affinity antibodies elution will take too long or require concentrations of competing peptide which render the method impracticable. Again the possibilities should be tested in small-scale experiments. Mix the beads with peptide for several hours then pellet by centrifugation and test the beads and supernatant for antigen complexes.

Protocol 10. Affinity precipitation of protein complexes with GST–fusion protein beads

Binding cell lysates labelled *in vivo* to GST–fusion protein beads is essentially performed as described for immunoprecipitation and immunopurification.

1. Add cell lysate to beads coated with GST–fusion proteins (*Protocol 1*) and rock for 1 h at 4°C for binding studies. If high backgrounds are encountered it may be necessary to preclear the lysate with beads coated with protein from an *E. coli* lysate expressing glutathione transferase leader protein from the pGEX parent plasmid.

2. Wash five times with lysis buffer.

3. Add 50 μl Laemmli loading buffer (see *Protocol 7*), heat, and run on SDS–PAGE.

Protocol 11. Affinity purification of GST–fusion protein or protein complexes

1. Prepare 1–2 ml of GST–fusion protein coated beads.

2. Mix with the cell lysate, 15 min at 4°C.

3. Wash 3 times with lysis buffer then elute the fusion protein and complexed proteins by competition with 1 bed volume of 50 mM Tris–HCl, pH 8.0, containing 5 mM reduced glutathione (Sigma, final pH 7.5, freshly prepared). Pellet the beads, repeat the elution, and pool the washes.

---

7. Separating proteins by polyacrylamide gel electrophoresis

The most convenient method of analysing protein complexes derived from the procedures described in this chapter is to separate them on polyacrylamide gels. Gel recipes are usually based on those of Laemmli (18) who originally described a discontinuous SDS buffer system which 'stacks' the protein sample during electrophoresis resulting in a ladder of sharply defined individual protein bands. The protein bands are visualized either by Coomassie blue or silver staining, or where lysates have been radiolabelled the gel can be soaked in fluorography reagents, e.g. Amplify (Amersham), and exposed to X-ray film. A general reference to these procedures is provided in reference 15.

8. Affinity blotting

An extension of polyacrylamide gel fractionation is to transfer unlabelled separated proteins to a nitrocellulose or nylon membrane by electroblotting where they can be probed directly for binding to antibodies (Western blotting) or to any other ligand. This technique can be adapted to look at the interaction of a specific protein probe with fractionated cell lysates, etc. This method has proved useful for studies on membrane proteins, particularly cell surface receptors which are often difficult to assay in cell extracts. For example, the epidermal growth factor (EGF) receptor from A431 human epidermoid carcinoma cell membranes was detected using mouse EGF and an anti-EGF antiserum labelled with [^{125}I]protein A (19). The technique relies on the ability to refold the target protein sufficiently to restore at least some of its binding activity. The gel separation can be performed using the Laemmli system described above, however, most proteins are completely unfolded when treated with SDS and sulfydryl reagents, forming randomly coiled polypeptide chains. Most of the denaturants are removed during blotting on to the membrane but not all proteins will recover biological activity, and it may be necessary to omit reducing agents (e.g. dithiothreitol, DTT) during the procedure or to separate the protein lysate using a nondenaturing gel (20).

Essentially, once the proteins are transferred from the separating gel to the membrane by electroblotting (15), the membrane is incubated with a buffer containing BSA or nonfat milk powder which serve to block nonspecific binding sites. The filter can then be incubated with the protein probe which has been tagged for detection by radiolabelling or using one of the other procedures described in Section 3. Alternatively, the bound protein can be detected using a specific antibody.

9. Direct probing of cDNA libraries with purified proteins

Expression libraries can be screened directly with labelled purified protein in much the same way as with antibodies. Plated phage are allowed to form plaques, which are transferred to nitrocellulose or similar filters under conditions which induce protein expression. Subsequently the filter is incubated directly with the labelled probe, or is first subjected to a denaturation/renaturation step before adding the probe. This approach has been successful in several cases, e.g. biotinylated Jun protein was used to detect several Jun binding proteins in λgt11 cDNA library derived from serum stimulated rat fibroblasts (9). Binding of the protein to the filter was detected in this case using streptavidin–alkaline phosphatase. Similarly, the Myc binding protein Max was obtained from a λgt11 cDNA library derived from a baboon lymphoblastoid cell line using an affinity purified GST–Myc fusion protein directly labelled with ^{125}iodine (21). Blanar and Rutter (10) used their *in vitro* phosphorylated c-Fos probe to find an associated helix–loop–helix zipper transcription factor, while Defeo-Jones *et al.* used anti-RB monoclonal antibodies to detect binding of purified RB protein to RB-binding proteins in a human fibroblast cDNA library (13). Positive plaques detected in the first round of screening should be picked into phage buffer, replated and screened again at a lower density until single positive plaques can be isolated. Confirmation that the phage-expressed cDNAs encode proteins which associate with the protein probe is most easily achieved by transferring the cDNA to a vector, e.g. one of the pGEM series (Promega) which can be used to prepare *in vitro* transcribed mRNA which is subsequently translated *in vitro* using rabbit reticulocyte lysate, in the presence of ^{35}S-labelled methionine. The lysate can be used to show protein binding in co-immunoprecipitation or affinity precipitation experiments with GST–fusion protein coated beads.

At this stage it is useful to design experiments using mutations or deletions in either the probe protein or the candidate binding protein, or to introduce competition for binding via other proteins or peptides (21).

Protocol 12. Screening an expression library for binding to labelled protein

Reagents

- TBST buffer: 10 mM Tris–HCl (pH 8.0), 150 mM NaCl, 0.05% Tween 20
- HBB: 20 mM Hepes (pH 7.2), 50 mM NaCl, 5 mM $MgCl_2$, 1 mM KCl, 5 mM DTT, 0.1% NP-40
- PBST: 137 mM NaCl, 2.76 mM KCl, 4.3 mM Na_2HPO_4, 0.2% Triton X-100

Method

1. Plate the λgt11 library on the Y1090 hsdR *E. coli* strain at a maximum density of 2×10^4/phage per 100 mm plate or 5×10^4/150 mm plate. It's important that the plaques are not overcrowded.

2. Incubate at 42 °C for 3–4 h. Meanwhile prepare the appropriate number of nitrocellulose filters by soaking them for a few minutes in 10 mM IPTG inducer and allow them to dry at room temperature.

3. Quickly overlay the plates with the filters and immediately transfer to 37 °C. Incubate for 4 h to overnight. Duplicate sets of filters can be lifted consecutively from the same plates for screening in duplicate to help eliminate false-positives.

4. Mark the filters by pushing a wide gauge needle right through the filter and the agar in at least three places. This will allow you to orientate the filters later.

5. Peel off the filters and transfer immediately to a large volume of TBST buffer. Shake gently for 10 min at room temperature to remove debris. Repeat this rinse twice with fresh buffer.

6. Carry out all steps at 4 °C from now on. You can *either* transfer the filters directly to HBB made 5% in low-fat dried milk, and block for an hour at 4 °C shaking gently.

7. *Or* you can subject the filter to a denaturation/renaturation step (22); submerge the filters in HBB buffer with 6 M guanidine–HCl, shake gently at 4 °C for 10 min. Remove the solution and repeat with fresh buffer. Shake for 10 min in HBB with 3 M guanidine–HCl and repeat three times with fresh buffer. Finally, block in HBB with 5% dried milk. Other blocking agents to try out are 5% BSA, or 20% fetal calf serum.

8. Transfer the filters to HBB with 1% milk after the blocking step, and incubate for 4 h to overnight with the probe protein (0.1–1 μg/ml). If high background is a problem it may help to also preblock the protein probe with a small volume of blocking buffer for an hour before adding it to the filters.

Protocol 12. *Continued*

9. Wash the filters 4 times for 10 min at room temperature in PBST at the end of the incubation. For [125]iodine-labelled filters, blot dry, wrap in clingfilm and expose to autoradiography film, otherwise develop filter by the appropriate method.

---

10. Expressing proteins by *in vitro* translation

Screening an expression library either directly with your labelled protein of interest or using an antibody raised against biochemically purified binding proteins will provide you with either complete or partial cDNAs encoding putative binding proteins. The next step is to look directly, by affinity precipitation or co-immunoprecipitation, at the interaction between your protein of interest and the proteins encoded by these cDNAs. The most convenient route to expression of putative binding proteins is to prepare large amounts of mRNA *in vitro* (23) which is achieved by cloning the cDNA downstream of a bacteriophage promoter, e.g. the SP6, T7, or T3 promoters. Convenient vectors for this purpose are available from several commercial sources, e.g. the pGEM series from Promega. The template plasmid is then transcribed *in vitro* by the appropriate specific bacteriophage DNA-dependent RNA polymerase and the resulting mRNA can then be efficiently translated in wheatgerm or rabbit reticulocyte extracts. Remember that if you only have a partial cDNA you will need to provide an in-frame ATG start codon during subcloning into the plasmid. The in-frame sequence around the *Eco*R1 cloning site of λgt11 is:

$$3' \ldots \text{GTC GAC CTT AAG GCG GCT ATG} \ldots 5'$$

It is usual to incorporate a 5′ terminal cap into the mRNA by including an excess of a cap analogue, e.g. G(5′)ppp(5′)G or one of its methylated derivatives in the transcription reaction. The cap structure stabilizes the mRNA and often enhances the efficiency of translation. [^{35}S]Methionine is included in the reaction so that the protein obtained can be subjected to co-immunoprecipitation, etc. as previously described.

Protocol 13. Preparation of mRNA *in vitro*

Reagents

- TE: 10 mM Tris–HCl (pH 7.4), 1 mM EDTA
- 10 × transcription buffer: 400 mM Tris–HCl (pH 7.5 at 37°C), 60 mM MgCl$_2$, 20 mM spermidine–HCl, 50 mM NaCl

Method

1. All solutions must be RNAase-free. Prepare 5 µg of linearized template plasmid DNA by cutting to completion with an enzyme which cuts once at the end of the cDNA to leave a protruding 5′ terminus, e.g. *Eco*R1, *Hind*111, or remove protruding 3′ ends using the 3′,5′-exonuclease activity of Klenow DNA polymerase.

2. Phenol:chloroform extract the template DNA twice and ethanol precipitate. Pellet the DNA and resuspend in TE at 1 µg/µl.

3. Mix the following components in the order shown. Warm everything to room temperature before use to avoid precipitating the DNA in the high concentration of spermidine:

   - 10 × transcription buffer   5 µl
   - H_2O 11 µl
   - 100 mM DTT   5 µl
   - 2.5 mg/ml BSA   2.5 µl
   - 5 mM rATP   2.5 µl
   - 5 mM rUTP   2.5 µl
   - 5 mM rCTP   2.5 µl
   - 0.5 mM rGTP   2.5 µl
   - 5 mM cap analogue   5 µl
   - DNA template (5 µg)   5 µl
   - 10 units/ml Human placental RNAase inhibitor   2.5 µl
   - 10–15 units/µl bacteriophage-encoded DNA-
     dependent RNA polymerase   3 µl

4. Mix by gently tapping the tube, and incubate for 1 hour at 37°C for bacteriophage T3 or T7 polymerase or at 40°C for SP6 polymerase. You can add more enzyme at the end of this time and repeat the incubation.

5. Add RQ1 RNAase-free DNAase (Promega) to a concentration of 1 unit/ µg of DNA template and incubate at 37°C for 15 min.

6. Extract the RNA once with an equal volume of phenol:chloroform and once with an equal volume of chloroform. Transfer the aqueous phase to a fresh tube and precipitate the RNA with 0.25 volume of 5 M ammonium acetate (Na^+ ions can inhibit translation) and 3 volumes of ethanol. Place at −20°C for 30 min.

7. Finally pellet the RNA by centrifugation at 12 000 *g* for 30 min, leave the tube open to dry on the bench and dissolve to about 1 µg/µl in TE (an OD_{260} of 1 is equivalent to an RNA solution of 40 µg/ml). You should check the size and integrity of the RNA on an agarose or acrylamide gel using RNA markers of known size.

10.1 Translation of synthetic mRNAs

Methods of preparing and using rabbit reticulocyte lysate are described elsewhere (24) and convenient kits are now available from several sources, e.g. New England Nuclear and Promega. The Promega system is already optimized for Mg^{2+} and K^+ concentration to work with the 'average' mRNA, but in other cases these ions should be titrated for translation efficiency with your RNA. Usually 70 mM KCl is optimal for translation of capped mRNA (try the range 20–80 mM) while magnesium acetate should titrate in the range 0–1 mM. The lysate should be thawed on ice just before use (unused lysate can be refrozen immediately at −70°C). Heat the mRNA at 67°C for 10 min to remove any secondary structure and cool on ice. Assemble the reaction components as advised by the manufacturer, including about 1 μCi/μl [35S]methionine or another appropriate amino acid. Incubate the reaction at 30°C for one hour and determine incorporation by TCA precipitation and gel analysis.

11. Antibodies raised against putative binding proteins

Where a cDNA sequence is recovered by screening expression libraries it is usually relatively straightforward to express and purify sufficient protein from immunization, e.g. using GST-fusions. Otherwise proteins separated by SDS–PAGE often induce a good immune response (15). Alternatively, where affinity techniques are used to trap protein complexes on beads you may be able to immunize directly with the protein-coated beads.

Immunoassays can provide a sensitive assay for screening hybridoma supernatants where microtitre wells are coated with protein complexes or complex components (antibody capture assay). Once antibodies are available to more than one component of the complex, two-site immunoassay (or two antibody sandwich assay) can provide a lot of information about protein interactions. Here the microtitre plate is coated with one antibody and the antigen complex allowed to bind. Unbound proteins are removed by washing and a second labelled antibody to a different component is added. Subsequent measurement of the amount of bound second antibody can give information about the presence of the complex in test samples, while antigen competition experiments can be designed to measure the level of specific components and to study their interaction with other members of the complex. Bartek *et al.* used two-site immunoassays to measure levels of SV40 Large T antigen (Tag) complexed to RB protein, and also the fraction of RB that was complexed to cellular p53, in SV40-transformed cells. Since no complex of p53 with RB could be detected in uninfected cells, they were able to deduce that p53 is complexed to RB via Tag to form a p53–Tag–RB tripartite complex (25).

Table 3. ELISA buffers

A. Plate-coating buffer	0.1 M CO_3/HCO_3, pH 9.0–9.8	0.42 g NaCO$_3$ + 2.52 g NaHCO$_3$ to 340 ml with H_2O
B. Substrate buffer	0.1 M citrate/phosphate, pH 6.0	7.05 g Na$_2$HPO$_4$ + 3.15 g citric acid to 650 ml with H_2O
C. Substrate	3,3′,5-5′tetramethylbenzidine (TMB)	10 mg/ml TMB in DMSO

The success of this assay is dependent on the avidity of both antibodies and not all will be suitable. You will need to titrate dilution series of each antibody. The second antibody can be labelled with iodine, enzymes, or biotin. The following protocol describes a procedure for use with antiserum from different species (in this example a mouse monoclonal and a rabbit polyclonal serum) which allows you to detect the second antibody indirectly with a third enzyme-linked antibody.

Protocol 14. Two-site ELISA assay

1. Incubate a 96-well microtitre plate overnight at room temperature with 50 µl/well of 30 µg/ml pure monoclonal antibody or 50 µl/well of ascites fluid diluted about 1 in 500 in buffer A (see *Table 3*)

2. Rinse the plate once in PBS and block for 2 h at room temperature with 3% BSA in PBS.

3. Wash the plate once in PBS, twice in PBS with 0.1% NP-40, and once again in PBS.

4. Add the antigen as 50 µl/well of serial two-fold dilutions. Incubate for 2 h at 4°C.

5. Wash the plate as in step 3.

6. Add 50 µl/well of the second rabbit antiserum (or your directly labelled second monoclonal or polyclonal antibody where appropriate) diluted in PBS, 1% BSA. Incubate for 2 h at 4°C.

7. Wash plate as in step 3.

8. Add 50 µl/well peroxidase-conjugated swine antiserum to rabbit immunoglobulin (Dako, typically diluted 1 in a 1000 in PBS, 1% BSA). Incubate for 2 h at 4°C.

9. Wash as in step 3.

10. Visualize the bound peroxidase-linked reporter antibody with TMB. Mix 6 ml of buffer B with 60 µl of solution C (see *Table 3*) and 15 µl of

Protocol 14. *Continued*

hydrogen peroxide. Immediately add 50 μl/well, leave for up to 30 min for colour to develop and then stop the reaction with 50 μl/well 1 M H_2SO_4.

11. Use an ELISA plate reader to measure OD_{450}.

References

1. Nilsson, B. and Abrahmsen, L. (1990). In *Methods in enzymology* (ed. D. V. Goeddel), Vol. 185, p. 144. Academic Press, London.
2. Riggs, P. (1990). In *Current protocols in molecular biology* (ed. F. M. Ausebel) Supp. 19, Unit 16.6, p. 1. Greene Associates/Wiley Interscience, New York.
3. Smith, D. B. and Johnson, K. S. (1988). *Gene*, **67**, 31.
4. Stuber, D., Matile, H., and Garotta, G. (1990). In *Immunological methods* (ed. I. Lefkovits and B. Pernis), Vol. IV, pp. 121–52. Academic Press, New York.
5. Hopp, T. B., Prickett, K. S., Price, V. L., Libby, R. T., March, C. J., Cerretti, D. P., Urdal, D. L., and Conlon, P. J. (1988). *Biotechnology*, **6**, 1204.
6. Dreher, M. I., Gherardi, E., Skerra, A., and Milstein, C. (1990). *J. Immunol. Methods*, **139**, 197.
7. Kaelin, W. G., Pallas, D. C., DeCaprio, J. A., Kaye, F. J., and Livingston, D. M. (1990). *Cell*, **64**, 521.
8. Fraker, P. J. and Speck, J. C. (1978). *Biophys. Res. Comm.*, **80**, 849.
9. Macgregor, P. F., Abate, C., and Curran, T. (1990). *Oncogene*, **5**, 451.
10. Blanar, M. A. and Rutter, W. J. (1992). *Science*, **256**, 1014.
11. Bo-Liang, L., Langer, J. A., Schwartz, B., and Pestka, S. (1989). *Proc. Natl Acad. Sci.*, **86**, 558.
12. Kaelin, W. G., Krek, W., Sellers, W. R., DeCaprio, J. A., Ajchenbaum, F., Fuchs, C. S., Chittenden, T., Li, Y., Farnham, P. J., Blanar, M. A., Livingston, D. M., and Flemington, E. K. (1992). *Cell*, **70**, 351.
13. Defeo-Jones, D., Huang, P. S., Jones, R. E., Haskell, K. M., Vuocolo, G. A., Hanobi, M. G., Huber, H. E., and Oliff, A. (1991). *Nature*, **352**, 251.
14. Storrie, B. and Madden, E. A. (1990). In *Methods in enzymology* (ed. M. P. Deutscher), Vol. 182, pp. 203–25. Academic Press, London.
15. Harlow, E. and Lane, D. (1988). *Antibodies: a laboratory manual*. Cold Spring Harbor Laboratory Press, New York.
16. Barak, Y. and Oren, M. (1992). *EMBO J.*, **11**, 2115.
17. Clarke, C. F., Cheng, K., Frey, A. B., Stein, R., Hinds, P. W., and Levine, A. (1988). *Mol. Cell. Biol.*, **8**, 1206.
18. Laemmli, U. K. (1970). *Nature*, **277**, 680.
19. Fernandez-Pol, J. A. (1982). *FEBS Lett.*, **143**, 86.
20. Soutar, A. K. and Wade, D. P. (1990). In *Protein function: a practical approach* (ed. T. E. Creighton), p. 55. IRL Press, Oxford.
21. Blackwood, E. M. and Eisenman, R. N. (1991). *Science*, **251**, 1211.
22. Vinson, C. R., LaMarco, K. L., Johnson, P. F., Landshultz, W. H., and McKnight, S. L. (1988). *Genes Develop.*, **2**, 801.

23. Melton, D. A., Kreig, P. A., Rebagliati, M. R., Maniatis, T., Zinn, K., and Green, M. R. (1984). *Nucl. Acids Res.*, **12**, 7035.
24. Sambrook, J., Fritsch, E. F., and Maniatis, T. (1989). *Molecular cloning: a laboratory manual*. Cold Spring Harbor Laboratory Press, New York.
25. Bartek, J., Vojtesek, B., Grand, R. J. A., Gallimore, P. H., and Lane, D. P. (1992). *Oncogene,* **7,** 101.

7

Using the two-hybrid system to detect protein–protein interactions

PAUL L. BARTEL, CHENG-TING CHIEN,
ROLF STERNGLANZ, and STANLEY FIELDS

1. Introduction

Protein–protein interactions are essential in almost all biological processes, including replication, transcription, secretion, signal transduction, and metabolism. Thus a central question in the study of any protein is to determine what other proteins are in contact with it. The answer, in the case of oncogene-encoded proteins for example, has provided enormous insights into the problems of cell cycle control and differentiation. It is, therefore, clear that intense research efforts will continue to focus on identifying proteins that interact with some protein of interest, referred to here as the target protein.

Current approaches to detect interacting proteins include traditional methods such as co-immunoprecipitation, crosslinking, and copurification through gradients or chromatographic columns (see Midgley and Lane, Chapter 6 of this volume). These biochemical methods have the major disadvantage that interacting proteins are generally known only as bands of a particular relative mobility on a polyacrylamide gel. To progress from these bands to cloned genes is often a difficult undertaking, involving such methods as the initial purification of sufficient protein for amino acid sequencing or antibody production, followed by additional work to screen a library for the corresponding gene (see Simmons, Chapter 5 of this volume).

To circumvent these difficulties, newer methods have been developed that result in the immediate availability of the cloned genes for proteins interacting with a target protein. One particularly powerful approach is to probe expression libraries with labelled target protein, using this protein in a manner similar to antibody probing of λgt11 libraries (1). This approach has identified genes for proteins that interact with calmodulin (2), jun (3), myc (4), the EGF receptor (5), and the retinoblastoma protein (6). However, this approach also has certain inherent limitations. The protein interactions are detected *in vitro*, generally on nitrocellulose filters. Thus the interactions must be stable to the buffer conditions that are used. Many proteins may not have the correct *in vitro* properties (folding, solubility, phosphorylation, etc.) for this method. In addition, conditions may need to be adjusted for each individual target protein to optimize the signal to noise ratio.

2. The two-hybrid system

We have developed a genetic system to detect protein–protein interactions *in vivo* which allows the rapid identification of genes encoding proteins that interact with a target protein. This two-hybrid system is based on the modular nature of many eukaryotic transcriptional activators, such as the yeast GAL4 protein, and the results of a number of critical earlier experiments. First, results from several laboratories, particularly those of Ptashne with GAL4 and Struhl with the yeast GCN4 protein, established that these activators contain separable domains for DNA-binding and transcriptional activation (7–9). The DNA-binding domain functions to localize the transcription factor to specific DNA sequences present in the upstream region of genes that are regulated by this factor. The activation domain functions to contact other components of the transcription machinery in order to initiate the transcription process. Second, results with such activators as the herpes virus protein VP16 were interpreted to suggest that these proteins might not directly contact DNA (10). Instead, these activators were postulated to function by binding to other proteins that were in turn bound to specific DNA sites. This idea was strongly supported by an experiment (11) with the yeast GAL80 protein, which normally binds GAL4 and inhibits its activity. Insertion of an activating sequence into GAL80 converted it into an activator that does not bind DNA directly, but instead contacts a DNA-bound GAL4 derivative. Third, the results of Brent and Ptashne (12) demonstrated that a hybrid transcriptional activator could be generated. In this experiment, a fusion gene was constructed encoding a hybrid of the *Escherichia coli* repressor LexA with the yeast GAL4. This hybrid protein could activate transcription in yeast of a gene containing a LexA operator, indicating that the LexA portion conferred the site-specific DNA-binding and the GAL4 portion contributed the transcription activation function. This experiment demonstrated that for one class of proteins, namely transcriptional activators, the structure was sufficiently modular that totally different proteins could be cut and pasted together through recombinant DNA technology to generate a functional hybrid.

The feasibility of the two-hybrid approach was shown by Fields and Song (13) in which the interaction of two yeast proteins, SNF1 and SNF4, was used to reconstitute GAL4 activity. SNF1, a protein kinase, was fused to the DNA-binding domain of GAL4, and SNF4, a protein associated with this kinase, was fused to the activation domain of GAL4. While neither hybrid protein alone could activate transcription of a gene containing GAL4 sites, the presence of both hybrids in the same cell allowed the SNF1–SNF4 interaction to bring the two GAL4 domains into sufficient proximity for transcriptional activation. This experiment suggested that not only transcriptional activators, but many, if not most, proteins could be made into hybrid proteins. These hybrids would each possess two functions, one involved in transcription (either DNA-binding or activation) and the other involved in protein–protein contacts. Furthermore, this experiment supported the idea that many of these protein–protein interactions might allow transcription to occur. Thus it was suggested that this strategy could be used to search libraries of activation domain hybrids in order to identify new proteins binding to a target protein present as a hybrid with a DNA-binding domain.

A schematic diagram of how the two-hybrid system works is shown in *Figure 1*. Plasmids are constructed that encode two hybrid proteins: one consists of the DNA-binding domain of GAL4 fused to one test protein 'X' and the other consists of the GAL4 activation domain fused to another test protein 'Y'. These plasmids are transformed into a strain of the yeast *Saccharomyces cerevisiae* that contains a reporter gene (e.g. *lacZ*) whose regulatory region contains GAL4 binding sites. Either hybrid protein alone must be unable to activate transcription of the reporter gene, the DNA-binding domain hybrid because it does not provide activation function and the activation domain hybrid because it cannot localize to the GAL4 binding sites. Interaction of the two test proteins reconstitutes the function of GAL4 and results in expression of the reporter gene, which is detected by an assay for the reporter gene product.

Use of the two-hybrid system or related methodology has been extended to screen activation domain libraries for proteins that interact with a target protein (14, 15). Total genomic or cDNA sequences are fused to the DNA encoding an activation domain. This library and a plasmid encoding a hybrid of the target protein fused to the DNA-binding domain are cotransformed into a yeast reporter strain, and the resulting transformants are screened for those that express the reporter gene. These colonies are purified and the plasmids responsible for reporter gene expression are isolated. DNA sequencing is then used to identify the proteins encoded by the library plasmids.

The two-hybrid system has a number of advantages. The assay is performed *in vivo*, under conditions similar to those in which protein interactions normally occur. Purified target protein or antibody against this protein is not required to detect the interaction. When positive plasmids are identified by screening an activation domain library, a portion of the gene encoding the interacting protein is immediately available. The two-hybrid system also appears to be more sensitive than co-immunoprecipitation and, therefore, may detect low-affinity interactions. This sensitivity may be due to the ability to detect weak or transient interactions which result in the accumulation of a

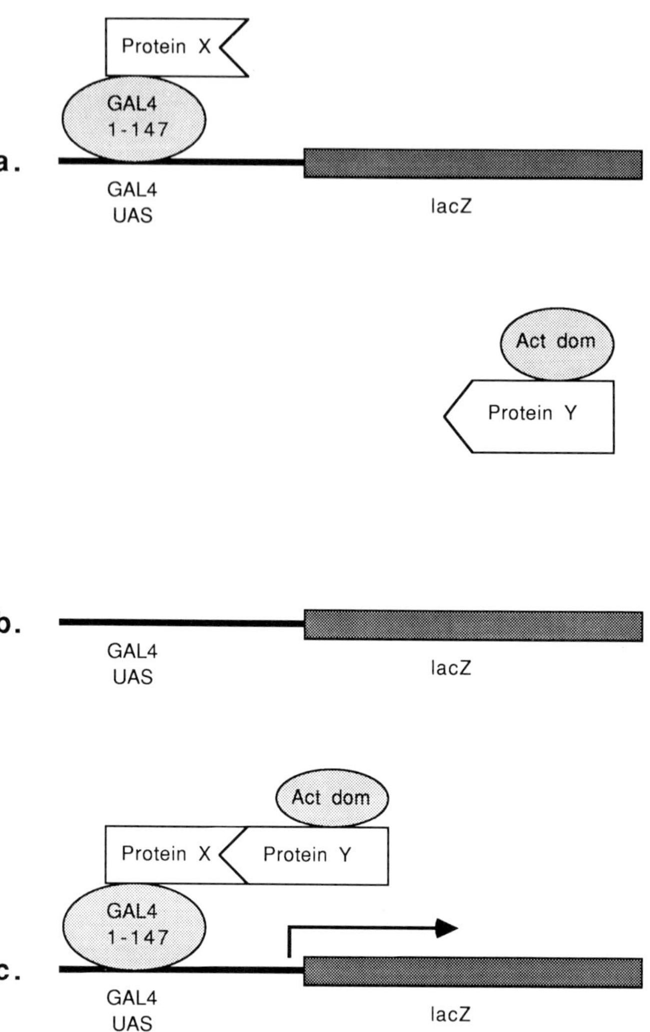

Figure 1. The two-hybrid system. (a) A hybrid consisting of the GAL4 DNA-binding domain fused to a protein X is localized to the GAL UAS upstream of the *lacZ* reporter gene. It is unable to activate transcription because it lacks a transcriptional activation domain. (b) A hybrid consisting of the GAL4 activation domain fused to a protein Y is unable to localize to the upstream region of the reporter gene. (c) A protein–protein interaction between X and Y brings the GAL4 domains into close proximity and reconstitutes transcriptional activity.

stable product such as β-galactosidase. This system can be used to map protein domains that are responsible for protein interactions and to identify mutations that affect these interactions. Disadvantages of the system are that the gene encoding the target protein must be available and that some hybrid proteins might not be stably expressed or localized to the nucleus. In addition, the GAL4 domains may interfere with the ability of the test proteins to interact.

The following sections of this chapter include protocols and background material needed to apply the two-hybrid system to detecting protein interactions between known proteins and between a target protein and library-encoded proteins. Also discussed are potential problems encountered when using the two-hybrid system. The chapter ends with a discussion of other uses for this system.

3. Testing for an interaction between known proteins

3.1 Introduction

Often, two proteins can be assigned to a common pathway based on genetic or biochemical evidence, or by homology to other proteins. If the genes encoding both proteins have been cloned, the two-hybrid system can be used to assay for a possible protein–protein interaction. Gene fusions are generated that encode two hybrid proteins. One of the candidate proteins (X) is fused to the GAL4 DNA-binding domain and the other protein (Y) is fused to the GAL4 activation domain. The two plasmids encoding these hybrids are cotransformed into a yeast reporter strain, and transformants are assayed for the production of β-galactosidase. The expression of β-galactosidase indicates that the two hybrid proteins interact and reconstitute GAL4 function.

Since the interaction between SNF1 and SNF4 was demonstrated, interactions between many pairs of proteins have been detected by constructing hybrids, some of which are listed in *Table 1*. Many of the interactions detected have been between two different proteins, i.e. heterodimers, but homodimers have also been detected. Examples of proteins that form detectable homodimers include yeast SIR4 and HIV gag.

Table 1. Some protein interactions detected by the two-hybrid system

Protein X	Protein Y	Reference
SNF1	SNF4	13
SIR4	SIR4	14
c-jun	c-fos	16
R6K rep	R6K rep	17
maize B	maize C1	18
HIV gag	HIV gag	34
HIV proteinase	HIV proteinase	G. Zybarth and C. Carter (personal communication)

3.2 Experimental procedure

Hybrid proteins can be constructed with any of a variety of DNA-binding domains or activation domains, but for simplicity we will limit this discussion to the use of the GAL4 DNA-binding domain (amino acids 1–147) and the GAL4 activation domain (amino acids 768–881). Vectors employing other domains are available and give similar results; the method appears to work with any DNA-binding and activation domains, which need not come from the same transcription factor. The basic procedure for testing two proteins for an interaction by the two-hybrid system is provided in *Protocol 1* and is discussed in greater detail below.

The two hybrid proteins are expressed at high levels from separate plasmids. Both hybrids are targeted to the yeast nucleus by nuclear localization sequences that are either an intrinsic part of the GAL4 DNA-binding domain (19) or have been added to the activation domain. The vectors have unique restriction sites located at the 3′ end of the open reading frame for either the DNA-binding or activation domain. The gene encoding the protein of interest is ligated into one of these sites in the correct orientation and with the correct reading frame such that a hybrid protein is expressed.

The two plasmids encoding the hybrid proteins are cotransformed into the yeast reporter strain along with all pairwise combinations of vectors and hybrid constructions as controls. The cells are plated on selective media to ensure that the colonies contain both plasmids. After the transformants have grown, they are assayed for expression of the chromosomally integrated reporter gene (β-galactosidase assay). If the two target proteins interact, GAL4 transcriptional activation function will be reconstituted and the transformants will express the *lacZ* gene. The control strains, which carry all other pairwise combinations of DNA-binding domain and activation domain plasmids, should not produce β-galactosidase activity. When several proteins are believed to operate in the same pathway, hybrids of all of these can be constructed and all pairwise combinations can be tested for possible interactions. Interactions that are identified can be the basis for further studies of the domains or residues involved in the interaction.

If the DNA-binding domain hybrid results in expression of the reporter gene in the absence of an activation domain hybrid, it suggests that the target protein contains a transcriptional activation domain. The test proteins can then be switched to the other vector (from the DNA-binding domain to the activation domain vector and vice versa). Alternatively, deletions of the gene can be constructed to assay for those that no longer have activation function; however, such deletions might also eliminate a potential interaction domain.

Protocol 1. Testing for an interaction between two known proteins

1. Construct two fusion genes. One is a fusion between the GAL4 DNA-binding domain and the gene encoding some protein X. The other is a fusion between the GAL4 transcription activation domain and some protein Y. The orientation and reading frame between the two portions of each fusion must be maintained so that hybrid proteins consisting of protein X fused to the DNA-binding domain and protein Y fused to the activation domain will be expressed (see *Protocol 2*).

2. Cotransform the yeast reporter strain with all pairwise combinations of vectors and gene fusion constructs (see *Protocol 3*). Plate the transformed cells on the appropriate selective media and incubate at 30°C.

3. After colonies appear (about 2–4 days), test the transformants for β-galactosidase activity. The transformants can be replica-plated to selective X-gal plates which are then incubated at 30°C until blue colonies appear, or they can be tested directly using the filter assay (see *Protocol 4*). If the β-galactosidase activity needs to be quantified, use the liquid assay (see *Protocol 5*).

4. The transformants expressing both relevant hybrids will turn blue (on X-gal) if a protein interaction is detected. All other combinations of plasmids should remain white.

3.3 Vectors and strains for the two-hybrid system

A number of vectors for the two-hybrid system are available. These plasmids are shuttle vectors which replicate autonomously in both *E. coli* and *S. cerevisiae*. They also carry selectable markers, the *bla* gene which confers ampicillin resistance in *E. coli*. and one of several nutritional genes (*TRP1*, *HIS3*, *URA3*, *LEU2*) that allow yeast auxotrophs to grow on selective media. Two vectors are required, one for the DNA-binding domain hybrid, and one for the activation domain hybrid. The reporter strain must contain mutations in the nutritional genes carried by the two vectors, and must contain a reporter gene that is transcriptionally activated through upstream binding sites corresponding to the DNA-binding domain. Some available vectors and strains are discussed in more detail below and are listed in *Table 2*.

These vectors often use the constitutive *ADH1* promoter to drive transcription of the hybrid gene, although newer vectors using a regulatable promoter will make it easier to test positives for dependence on both hybrids and will eliminate some problems associated with expressing toxic proteins. *Figure 2* shows the restriction maps of two useful DNA-binding domain vectors, pMA424 and pGBT9.

Activation domain vectors are used to express hybrids with both known proteins and proteins that are encoded by cDNA or genomic DNA libraries. Although the GAL4 activation domain does not normally contain a nuclear localization sequence, the SV40 large T-antigen nuclear localization sequence has been engineered on to its N-terminus (14). The GAL4 activation domain, which normally functions at the C-terminus of the protein, also functions

Table 2. Vectors and strains used in the two-hybrid system

Plasmid	Characteristics	Reference
pMA424	GAL4 DNA-binding domain, *HIS3*, ~12 kb EcoRI, BamHI, SalI sites	20
pBTM116	*lexA, TRP1*, 5.4 kb EcoRI, SmaI, BamHI, SalI, PstI sites	Bartel and Fields
pGBT9	GAL4 DNA-binding domain, *TRP1*, 5.4 kb EcoRI, SmaI, BamHI, SalI, PstI sites	Bartel and Fields
Activation domain vectors		
pGAD2F	GAL4 activation domain, *LEU2, ~13 kb* BamHI site[a]	14
pGAD10	GAL4 activation domain, *LEU2*, 6.6 kb BglII, XhoI, BamHI, EcoRI sites	Bartel and Fields
pGAD424	GAL4 activation domain, *LEU2*, 6.6 kb EcoRI, SmaI, BamHI, SalI, PstI sites	Roeklein and Fields
pSE1107	GAL4 activation domain, *LEU2*, BglII, BamHI, XhoI sites	Elledge
pSD.10	VP16 activation domain, *URA3*, EcoRI, BstXI, XhoI	15

[a] Variants of this plasmid, namely pGAD1F and pGAD3F, contain the BamHI site in the other two reading frames relative to the activation domain.

Reporter strains

Strain	Full genotype	Reporter	Transformation markers	Reference
GGY1::171	*MATα leu2-3,112 his3-200 met⁻ tyr1 ura3-52 ade2 gal4Δ gal80Δ URA3::GAL1-lacZ*	*GAL1-lacZ*	*his3, leu2*	21
Y153	*MATa ura3-52 leu2-3,112 his3-200 ade2-101 trp1-901 gal4Δ gal80Δ LYS2::GAL1-HIS3 GAL1::GAL1-lacZ GAL1-lacZ* (unknown location)	*GAL1-HIS3, GAL1-lacZ*	*trp1, leu2*	Elledge
CTY1	*MATα ade⁻ ura3-52 his3-200 leu2-3,112 lys2-801 trp1-901 gal4⁻ gal80⁻ URA3::GAL1-lacZ lys2::GAL1-HIS3*	*GAL1-lacZ, GAL1-HIS3*	*his3,leu2,trp1*	Chien and Sternglanz
CTY10-5d	*MATa ade2 trp1-901 leu2-3,112 his3-200 gal4⁻ gal80⁻ URA3::lexA-lacZ*	*lexA-lacZ*	*his3,trp1,leu2*	Chien and Sternglanz
YPB2	*MATa ura3-52 his3-200 ade2-101 lys2-801 trp1-901 leu2-3,112 can^R gal4-542 gal80-538 LYS2::GAL1-HIS3 URA3::(GAL 17mers)-lacZ*	*GAL1-HIS3* (GAL 17mers) *-lacZ*	*trp1,leu2*	Bartel and Fields

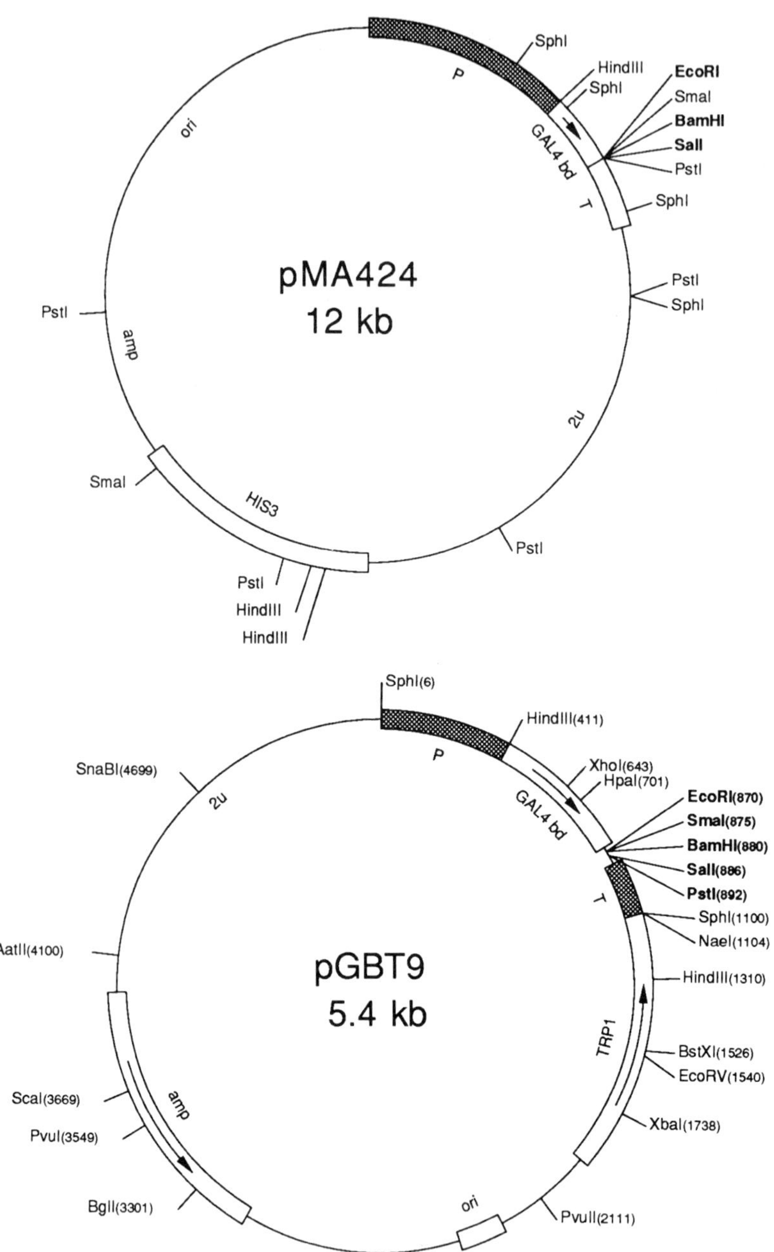

Figure 2. DNA-binding domain vectors pMA424 (reference 20) and pGBT9. Important features include the *ADH1* promoter (P) which drives transcription of the GAL4 DNA-binding domain (GAL4 bd), and the *ADH1* transcriptional terminator (T). Restriction sites in bold are unique and suitable for insertion of heterologous genes. Nucleotide positions of the restriction sites for pGBT9 are shown.

when placed at the N-terminus of a hybrid protein. Multiple restriction sites are placed at the 3′ end of the open reading frame encoding the activation domain to allow the construction of a fusion gene that expresses a hybrid of the activation domain with some protein. Maps of two activation domain vectors, pGAD2F and pGAD424 are shown in *Figure 3*.

Depending on the DNA-binding domain and activation domain vectors that are used, different reporter strains can be transformed with these plasmids. The most commonly used reporter is the bacterial *lacZ* gene which encodes β-galactosidase, but selectable genes, such as *HIS3*, are also used. The reporter genes contain upstream binding sites for the specific DNA-binding domain. When using the GAL4 DNA-binding domain, the strain must be *gal4* so that endogenous GAL4 does not activate transcription of the reporter gene. It is helpful to use *gal180* strains when using the GAL4 activation domain to avoid the need to add galactose to the growth medium.

3.4 Construction of the hybrid proteins

Construction of hybrid genes encoding hybrid proteins requires that in-frame fusions to the DNA-binding or activating domain are made in the appropriate vector. Sometimes, a fusion gene can be easily generated if compatible restriction sites are present in the test genes and one of the vectors. If not, the gene fragment can be generated by PCR with useful restriction sites incorporated into the primers. Often a restriction site can be changed into a different site or put into a different reading frame by incorporation of a short adapter oligonucleotide.

Generation of a fusion gene is performed using standard molecular biology techniques. A short outline of the steps for making a fusion are given in *Protocol 2*. Constructions can be checked by restriction site or sequence analysis. A useful sequencing primer for reading from the GAL4 DNA-binding domain junction corresponds to amino acid residues 132 to 137: 5′-TCA TCG GAA GAG AGT AG-3′, and from the GAL4 activation domain junction corresponds to amino acid residues 758 to 763: 5′-TAC CAC TAC AAT GGA TG-3′.

Protocol 2. Construction of hybrid genes

1. Purify the gene fragment, whether generated by PCR or cut out of some available plasmid, by any of several standard methods (22).

2. Digest the DNA-binding or activation domain vector with the appropriate restriction enzyme(s), treat with phosphatase, and purify.

3. Prepare ligation mixtures of the vector and insert. After incubation introduce them into *E. coli*.

4. Identify the correct clones by restriction analysis of representative plasmids.

Protocol 2. *Continued*

5. Check the orientation and framing by sequencing across the junction between the genes encoding the DNA-binding domain and protein X or between the genes encoding the activation domain and protein Y.

3.5 Growth and maintenance of yeast

Recipes for the different types of media needed for use in the two-hybrid system are presented in *Table 3*. For additional information about growing and maintaining yeast see reference 23. Most strains of yeast can be propagated on YEPD plates, which are incubated at 30°C for 1–2 days until colonies appear. Stocks of yeast strains can be stored in 15% glycerol at −70°C. A small amount of frozen cells can be scraped from this stock with a loop or wooden stick and streaked on to YEPD plates to recover the strain. YEPD media without agar is used to grow yeast strains for transformations (see *Protocol 3*). Synthetic media are prepared to select for yeast cells that have been transformed with activation domain and DNA-binding domain plasmids.

Synthetic media consists of yeast nitrogen base, a carbon source (generally at 2% concentration), and a stock of dropout solution. Dropout solutions contain essential nutrients, such as amino acids and nucleotides, with one or more of them omitted. Thus only transformants carrying the appropriate nutritional genes can grow in these synthetic media. For instance, a 10 × -leu stock would be used to prepare media that select for transformants carrying a *LEU2*-bearing vector. Glucose is most commonly used as the carbon source, but because it causes repression of transcription at the *GAL1,10* promoter, reporter strains that use this promoter should be grown on a nonrepressing carbon source such as sucrose, raffinose, or a combination of glycerol, ethanol, and galactose. X-Gal media, used to assay yeast colonies for the production of β-galactosidase, is also prepared using the appropriate dropout solution to maintain selection for plasmids. Yeast cells grow poorly on X-gal plates and, therefore, cannot be directly plated on this media following a transformation procedure. Instead, they must be replica-plated from another plate in order to assay for β-galactosidase.

3.6 Yeast transformation

Fusion genes are introduced into a yeast reporter strain by transformation. Plasmid DNA can be purified from *E. coli* using standard protocols (22), with mini-prep DNA sufficient for most uses. The use of Li$^+$ to prepare competent yeast cells, developed by Ito *et al.* (24), was modified by Schiestl and Gietz (25) to give a higher frequency of transformation. This modified method is outlined in *Protocol 3*. Two plasmids can be introduced simultaneously into the same strain, but the efficiency is then lower. Some laboratories prefer to transform the strain with each plasmid sequentially, which is recommended

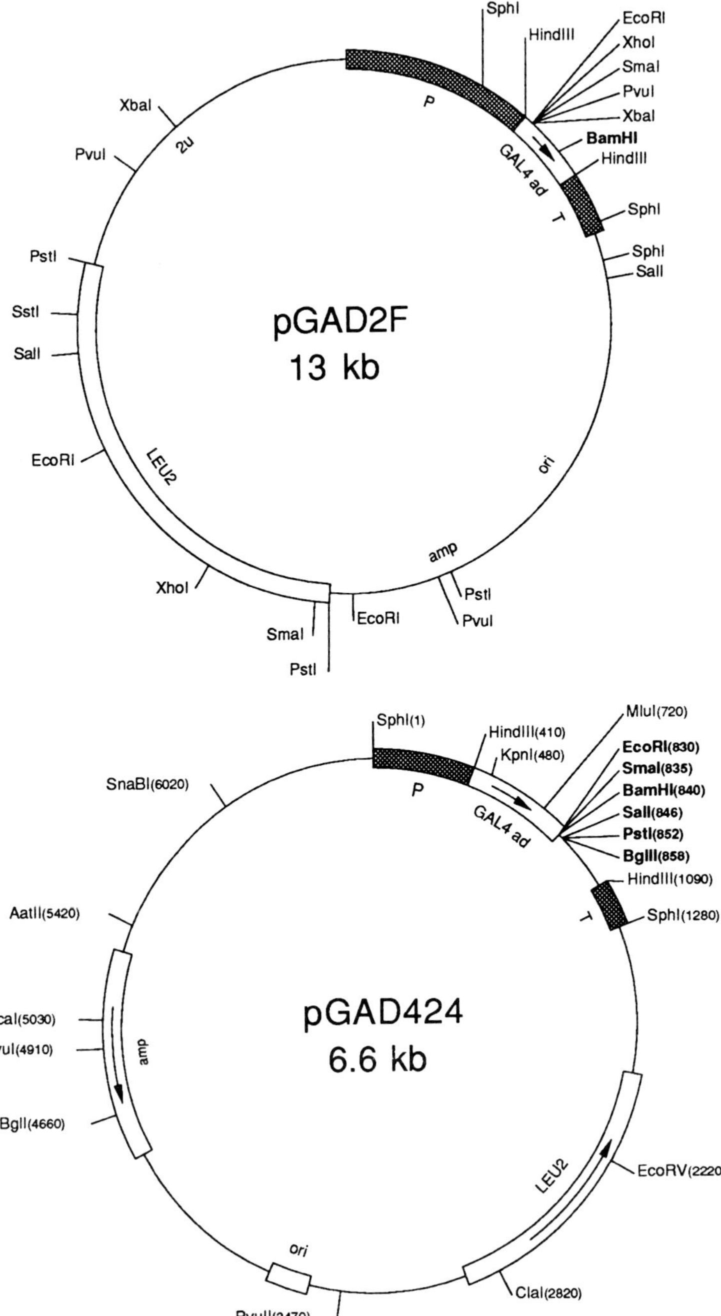

Figure 3. Activation domain vectors pGAD2F (reference 14) and pGAD424. Important features include the *ADH1* promoter (P) which drives transcription of the GAL4 activation domain (GAL4 ad), and the *ADH1* transcriptional terminator (T). Restriction sites in bold are unique and suitable for insertion of heterologous genes.

165

only if there is no selective disadvantage to cells expressing the first hybrid protein. In this case, the transformants can be grown in selective media or in YEPD (*Table 3*) before the second transformation; the cells lose plasmids at a low rate such that selective pressure is not absolutely required at this step.

Table 3. Recipes for yeast media

YEPD	Media used for the general maintenance of yeast strains	
	Difco peptone	20 g/l
	Yeast extract	10 g/l
	Agar (if needed for plates)	15 g/l

After autoclaving add glucose to 2% by adding 40 ml of a 50% solution to each litre of media

10 X Dropout solution	A solution of required amino acids and nucleotides that is added to synthetic media. Solutions lacking one or more of these components are used to make media to select for yeast transformants that contain plasmids with specific nutritional markers	
	L-Isoleucine	300 mg/l
	L-Valine	1500 mg/l
	L-Adenine hemisulfate salt	200 mg/l
	L-Arginine HCl	200 mg/l
	L-Histidine HCl monohydrate	200 mg/l
	L-Leucine	1000 mg/l
	L-Lysine HCl	300 mg/l
	L-Methionine	200 mg/l
	L-Phenylalanine	500 mg/l
	L-Threonine	2000 mg/l
	L-Tryptophan	200 mg/l
	L-Tyrosine	300 mg/l
	L-Uracil	200 mg/l

SD media	Synthetic media used to select for cells that are transformed with plasmids carrying specific nutritional markers	
	Difco yeast nitrogen base without amino acids	6.7 g/l
	Agar	15 g/l

After autoclaving add 100 ml of the appropriate 10 × dropout solution per litre and add glucose (or other carbon source, e.g. for s-raffinose or s-galactose) to 2%.

X-Gal plates	Used to detect colonies that produce β-galactosidase	
	Difco yeast nitrogen base without amino acids	6.7 g/l
	Agar	15 g/l

After autoclaving add sucrose to 2%, 100 ml of the appropriate dropout solution, 100 ml of 1 M phosphate buffer (pH 7.0), and 40 mg X-gal per litre of media

---

Protocol 3. Transformation of yeast (from Schiestl and Gietz (25) and Hill *et al.* (26))

1. Inoculate cells from an overnight culture into 50 ml YEPD (*Table 3*) and incubate at 30°C with shaking. Typically, add 0.1–0.2 ml saturated culture in the evening to get an $OD_{600} = 1–2$ the next morning.

2. Transfer enough of this culture to 300 ml YEPD to produce an $OD_{600} = 0.2$ (about 20 ml). Incubate at 30°C with shaking for 3–4 more hours.

3. Pellet the cells in a GSA rotor (5000 r.p.m. (4000 g), 5 min, room temperature).

4. Discard the supernatant and resuspend the pellet in 10 ml H_2O. Transfer the cells to a 50 ml centrifuge tube.

5. Pellet the cells in an SS34 rotor (7000 r.p.m. (6000 g), 5 min, room temperature).

6. Discard the supernatant and resuspend the pellet in 1.5 ml sterile 1 × TE/LiAC (made fresh from stocks of 10 × TE [0.1 M Tris–HCl, 0.01 M EDTA, pH 7.5] and 10 × LiAC [1 M lithium acetate adjusted to pH 7.5 with dilute acetic acid]).

7. Add the plasmid DNAs (1–5 μg) and single-stranded carrier DNA (200 μg) to a microfuge tube. The carrier DNA is produced by dissolving salmon sperm DNA in TE, sonicating it to reduce its viscosity, extracting it with phenol/chloroform, and precipitating it with ethanol. The DNA is resuspended at a concentration of 10 mg/ml in TE, placed in a boiling water bath for 20 min, and immediately cooled on ice. The carrier DNA solution can be stored in aliquots at −20°C.

8. Add 200 μl of the yeast suspension to each microfuge tube.

9. Add 1.2 ml sterile PEG solution (40% PEG, 1 × TE, 1 × LiAC, made fresh from stocks of 50% PEG 4000, 10 × TE, 10 × LiAC) to each tube and incubate at 30°C for 30 min with shaking.

10. Add dimethyl sulfoxide to 10%. Mix gently.

11. Heat pulse for 15 min in a 42°C waterbath.

12. Chill on ice and centrifuge for 5 sec to pellet the cells.

13. Aspirate off the supernatant and resuspend the cells in 1 ml TE. Plate 0.1 ml on to each standard size (100 mm) plate of the appropriate yeast dropout media. Plate 0.3 ml on to 150 mm plates.

14. Incubate the plates at 30°C until colonies appear, usually 2–3 days.

3.7 Assays for β-galactosidase

Transformants are allowed to grow at 30°C, usually for 2–4 days, until they are large enough to assay for β-galactosidase activity. Three different assays can be used to detect or quantitate β-galactosidase activity. In the first, colonies are replica-plated to X-gal plates (*Table 3*) lacking the appropriate nutrients, and incubated at 30°C. The colonies will grow slightly but they will not be very healthy, because of the phosphate buffer in the media. Positive colonies will turn blue at anytime from one to ten days. A second method is the filter assay (*Protocol 4*) which is faster and more sensitive (27). A sterile filter is layered over the transformants and lifted to remove cells that stick to the filter. The cells are permeabilized by freezing the filter in liquid nitrogen and thawing at room temperature. The filter is then layered on to another filter that is soaked with buffer containing X-gal. The filters are incubated at 30°C and checked for the appearance of blue colonies, which occurs anytime from one hour to one day. Positives can be isolated by picking the corresponding colonies from the original plates to fresh media. A similar method, used by Dalton and Treisman (15), is a modification of the filter assay. Transformed cells are plated directly on to sterile Hy-bond filters that have been layered on to selective synthetic media. After the transformants are grown, the filters are assayed as described above. Positive transformants are picked directly from the filter on to fresh media. A third method assays β-galactosidase activity in liquid culture (*Protocol 5*). The assay measures the release of *o*-nitrophenol, which has a yellow colour, from the substrate ONPG. The results are expressed as units, defined by Miller (28), which are calculated using the equation provided.

Protocol 4. Filter assay for the detection of colonies producing β-galactosidase

1. Prepare the following stock solutions:

   Z buffer without β-mercaptoethanol: per litre:
   - $Na_2HPO_4 \cdot 7H_2O$ 16.1 g
   - $NaH_2PO_4 \cdot H_2O$ 5.5 g
   - KCl   0.75 g
   - $MgSO_4 \cdot 7H_2O$ 0.246 g
   - Adjust pH to 7.0

   X-gal: dissolve 5-bromo-4-chloro-3-indolyl-β-D-galactoside (X-gal) in *N*,*N*-dimethylformamide at a concentration of 20 mg/ml

2. Prepare an X-gal/Z buffer solution containing 100 ml of Z buffer, 0.27 ml of β-mercaptoethanol, and 1.67 ml X-gal stock. To assay colonies on a standard 100 mm diameter Petri dish, pipette 1.8 ml of this solution to a clean Petri dish and soak a 75 mm filter placed on to the solution. Use either Whatman No. 1 or VWR grade 413 paper filters. A larger 150 mm diameter Petri dish requires 5 ml of solution and a 12.5 cm filter.

3. Place a sterile Whatman No. 1 or VWR grade 413 filter on to a transformation plate after the colonies have grown (approx. 2–3 days) and orient it with the agar medium by using a needle to make holes in three or more asymmetric locations. Nitrocellulose filters can be used but they are prone to crack when frozen and are also more expensive than paper filters. Using forceps, transfer the filter, flipped so that the colonies are facing up, to a pool of liquid nitrogen.

4. After the filter has completely frozen (approx. 5 sec) remove it from the liquid nitrogen and allow it to thaw at room temperature.

5. Carefully layer the thawed filter, colony side up, on to the filter presoaked with X-gal/Z buffer solution. Be sure to exclude air bubbles. Colonies producing β-galactosidase will turn blue any time from 30 min to 30 hours. The β-galactosidase-producing colonies can be identified by aligning the filter using the orienting marks.

Protocol 5. Quantitation of β-galactosidase activity

1. Inoculate yeast transformants into the appropriate synthetic media containing a nonrepressing (for GAL4-dependent reporters) carbon source such as sucrose or raffinose. Incubate at 30°C until the culture reaches midlog phase ($OD_{600} \sim 1.0$). Read the absorbance of the culture at 600 nm (OD_{600}).

2. To a 1.5 ml microfuge tube, add 0.1 ml of culture and 0.7 ml Z buffer (containing β-mercaptoethanol; see *Protocol 4*).

3. Add 50 μl chloroform and 50 μl 0.1% SDS. Vortex for 30 sec.

4. Add 0.16 ml of an ONPG solution (4 mg/ml in 0.1 M phosphate buffer, pH 7.0). Mix well.

5. Incubate at 30°C for one hour, unless the yellow colour develops much more rapidly. Quench the reactions by adding 0.4 ml of 1 M Na_2CO_3.

6. Spin for 10 min to remove the cell debris and read the absorbance at 420 nm (OD_{420}).

7. Calculate the activity using the following formula:

$$\beta\text{-galactosidase units} = 1000 \times \frac{OD_{420}}{t \times V \times OD_{600}}$$

where: t = time of incubation in minutes,
V = volume of culture added to Z buffer in ml.

3.8 Potential problems

Although some protein pairs actually interact *in vivo*, this interaction may not be detected by the two-hybrid system for a variety of reasons. If high-level expression of one or both of the hybrid proteins is toxic to the cell, transformants cannot be grown. Sometimes truncation of one of the proteins will alleviate the toxicity and still allow the interaction to occur. Alternatively, the hybrid proteins may not be stably expressed in the cell, the added domains may occlude the site of interaction, or the hybrid protein may fold improperly. Different hybrid constructions might help, but in some cases there may be no way to detect an interaction using this system.

A more common problem occurs when the protein fused to the DNA-binding domain activates transcription. This activation is especially likely if the protein is a transcription factor, although other proteins that are not involved in the transcription process are capable of activating transcription. The two test proteins can be switched to the opposite vector so that the activating protein is fused to the activation domain instead of the DNA-binding domain. It may also be possible to remove the activating domain of the target protein. Less frequently an activation domain hybrid will activate transcription inappropriately, possibly because it contains a DNA-binding domain that recognizes a sequence present in the reporter gene construct. This problem is discussed in more detail in Section 4.7.

4. Screening fusion libraries for proteins that interact with a protein of interest

4.1 Introduction

The two-hybrid system allows one to screen a large number of candidate proteins (about 10^7, depending on the scale of the yeast transformation) for those that interact with a target protein. Genomic or cDNA libraries are built into activation domain vectors such that random proteins fused to an activation domain will be expressed. Plasmids encoding the target protein hybrid and the library of activation domain hybrids are cotransformed into a reporter strain. Interaction of an activation domain hybrid with the target protein reconstitutes transcriptional activation function, leading to expression of the reporter gene. Only rarely will a transformant be positive, as the vast majority of transformants do not express the reporter gene because their activation domain fusions do not bind to the target protein. After a positive clone has been identified, it is easy to sequence the gene corresponding to the interacting protein because it is present as an insert in the activation domain vector.

The utility of this method in identifying interacting proteins from an activation domain library was first demonstrated for the yeast SIR4 protein (14). SIR4, which is involved in the transcriptional silencing of the silence mating type loci, was demonstrated to form oligomers by the two-hybrid system. SIR4 was then used as the target to screen a yeast genomic DNA library in an activation domain vector. These experiments identified a SIR4 hybrid as a positive from about 200 000 transformants, demonstrating that genes for proteins that interact with a target protein can be isolated using this method. Other successful searches using the yeast library include the identification of RIF1, which binds to RAP1 (29) and a protein that binds to the SNF1 protein kinase (30). Dalton and Treisman (15) extended this technique to mammalian genes by screening an activation domain/human cDNA library for proteins that interact with serum response factor (SRF).

Presently, few activation domain libraries are available for use with the two-hybrid system. It is likely that in the near future libraries from a variety of organisms and tissues will be available. *Table 4* lists some libraries that have been constructed.

4.2 Experimental procedure

Screening fusion libraries for proteins that interact with a protein of interest requires that the proper DNA-binding domain and activation domain constructs have been made. Genomic or cDNA libraries will nearly always be built as fusions to an activation domain since random hybrids with any DNA-binding domain will produce a large percentage of hybrids that have transcriptional activity on their own (20). For this reason, the protein of interest should be produced as a hybrid with the DNA-binding domain. It is important to test the hybrid for transcription activation function in the appropriate reporter strain because if it activates even slightly, it will be difficult to reliably screen for colonies expressing higher levels of the reporter gene. If the hybrid does not activate, its stability in the reporter strain can be assayed by Western blot. If a known protein partner is available, it can be used to check whether an interaction is detectable in the two-hybrid system. A yeast gene can be assayed for complementation of a mutation in that gene, as many hybrid genes are capable of such complementation. *Protocol 6* outlines the steps needed to screen an activation domain library for proteins that interact with a protein of interest.

Protocol 6. Experimental plan: Screening libraries for interacting proteins

1. Construct a fusion of the gene encoding the target protein to the GAL4 DNA-binding domain using one of the available vectors (*Protocol 2*). The gene should be inserted in-frame and in the correct orientation such that a hybrid consisting of the target protein fused to the GAL4 DNA-binding domain will be expressed from this vector.

2. Test the hybrid for activation function in an appropriate reporter strain. If the hybrid activates transcription of the reporter gene, a library search cannot be performed. If possible, modify the hybrid so that it no longer activates.

Protocol 6. *Continued*

3. Cotransform the reporter strain with both the plasmid encoding the target protein hybrid and with an activation domain library. Plate the cells on selective media and incubate at 30°C.

4. Assay transformants for β-galactosidase activity using the filter assay (*Protocol 4*). Blue colonies are isolated by restreaking and assaying again for β-galactosidase activity.

5. Isolate plasmid DNA from the positives (see *Protocol 7*), and use this to transform *E. coli*. Library plasmids are isolated and reintroduced into yeast with and without the plasmid encoding the DNA-binding domain hybrid.

6. Assay these transformants to determine whether the DNA-binding domain hybrid is required to mediate activation. Positive plasmids can be checked for promoter specificity (see Section 4.7). Sequence the insert of positive plasmids.

Table 4. Activation domain libraries

Organism	Vector	Complexity	Reference
S. cerevisiae (genomic)	pGAD1,2,3	5×10^6	14
S. cerevisiae (cDNA)	pSE1107	10^8	Elledge
Human B cell (cDNA)	pSE1107	10^8	Elledge
Human HeLa (cDNA)	pSD.10	10^7	15

4.3 Construction of a hybrid protein library

Like any other library, the easiest way to acquire an activation domain library is to get one that has already been constructed by someone else. If a suitable one is not available, it can be readily constructed. Libraries from organisms with genomes that contain few or no introns, such as bacteria or yeast, can be prepared from genomic DNA; otherwise cDNA is used.

Activation domain libraries obtained from others usually need to be amplified in order to have enough plasmid DNA for the yeast transformations. If the library is sent as DNA, it is introduced into *E. coli* by electroporation and the cells are plated on selective media (e.g. LB + ampicillin). If the library is sent as *E. coli* transformants, the cells are plated directly such that the resulting colonies are nearly confluent. In either case, the cells are scraped from the plates and plasmid DNA is isolated using standard techniques (22). Growing the transformants on solid media instead of liquid culture minimizes over-amplification of some library plasmids at the expense of others.

Libraries of genomic DNA inserts can be constructed from random DNA fragments. The DNA can be sheared, repaired and ligated to adapters such that the fragments can be inserted into digested activation domain plasmid. Alternatively, DNA can be partially digested by frequent-cutting restriction enzymes (e.g. *Alu*I, *Rsa*I, *Sau*3A). DNA prepared by any of the methods above should be size fractionated such that small DNA fragments will not be cloned, producing useless fusions. Greater complexity can be achieved by cloning the DNA into sets of vectors (such as pGAD1, 2, 3) that have cloning sites in all three reading frames. This is not necessary when using mechanically sheared DNA. To ensure that almost every protein is represented as a hybrid with the activation domain, libraries should be made as large as possible. For instance, a yeast (genome size = 2×10^7 bp) library consisting of 10^6 clones will have, on average, a productive fusion every 120 bp. One must take into account that only one of every two inserts will be in the correct orientation and of these, only one in three is in the correct reading frame. Additionally, because of statistical considerations, to have a 99% chance of containing a fusion every 120 bp, a yeast library would need to be five times larger and, therefore, would contain 5×10^6 clones.

Libraries constructed from cDNAs face different constraints. Standard methods for producing cDNAs are available (22), including methods for producing cDNAs with different restriction sites at the 5' and 3' ends to allow directional cloning into vectors containing the appropriate restriction sites. This ensures that all cDNAs are cloned in the correct orientation, but only one in three will be in the correct reading frame relative to the activation domain gene. cDNA libraries have the disadvantage of representing only expressed genes, and low abundance mRNAs will be represented very infrequently in the library. A large number of independent clones is required to ensure that proteins encoded by low abundance transcripts are represented in the library. However, it is worth keeping in mind that with present technologies, it is convenient to screen no more than about 10^7 yeast transformants by the two-hybrid system.

4.4 Cotransformation of target hybrid and the library

A search for interacting proteins requires that the appropriate reporter strain must be cotransformed with both the activation domain library and the plasmid encoding the DNA-binding domain hybrid. The two plasmids can be introduced simultaneously or sequentially. When two plasmids are introduced simultaneously, the efficiency of transformation is considerably reduced (as compared to a single transformation), so relatively large amounts of transforming DNA are required. If the plasmids are introduced sequentially, the DNA-binding domain plasmid is introduced first, the resulting transformant is grown and then transformed with library DNA. Because the library DNA is introduced in a single transformation step rather than as part of a double transformation, less library DNA is needed to obtain the same number of double transformants. Sequential transformation can become problematic if expression of the DNA-binding domain hybrid is detrimental to the cell. Because transformants expressing the hybrid must be regrown for the second transformation, plasmids that delete sequences for the DNA-binding domain hybrid, and thus are less toxic, may accumulate.

4.5 Isolation of positive clones

The relatively rare transformant that produces β-galactosidase can be identified by two methods. Transformants can be replica-plated to X-gal plates (see *Table 3*) on which blue colonies will appear anytime from one day to one week. Alternatively, a filter assay for β-galactosidase activity can be performed on the colonies (*Protocol 4*). Positives should be picked to fresh media and should be rechecked to see that they maintain the correct phenotype. From a crowded plate, it can be difficult to isolate the cells that produce the blue colour. These can be purified by restreaking for isolated colonies and assaying for colonies that produce β-galactosidase.

After a positive colony has been identified and purified, the library plasmid can be purified from the yeast cells. A rapid procedure for isolating DNA from yeast cells is presented in *Protocol 7* (31). This DNA is not useful for restriction analysis or sequencing but it can be used to transform *E. coli*. The use of a *leuB⁻ E. coli* strain selects for transformants that carry the *LEU2*-bearing library plasmid on minimal media. Standard plasmid mini-prep procedures are used to isolate the library plasmid. These preparations are used to retransform the reporter strain both with and without the plasmid encoding the DNA-binding domain hybrid. Assays for β-galactosidase activity will identify plasmids that require the DNA-binding domain hybrid to turn colonies blue. Often more than one library plasmid will be present in each yeast cell, and the plasmid responsible for the blue colour must be identified by cloning in bacteria.

Protocol 7. Isolation of DNA from yeast (from Hoffman and Winston (31))

Reagents

- lysis solution: 2% Triton X-100, 1% SDS, 100 mM NaCl, 10% mM Tris (pH 8.0), 1.0 mM EDTA

Method

1. Inoculate the yeast transformant into 2 ml of media. Incubate at 30 °C until the culture is saturated.

2. Fill a 1.5 ml microfuge tube with culture and spin at 16 000 *g* in the

minifuge for 5 sec to pellet the cells. Decant the supernatant and re-suspend the pellet in the residual liquid by vortexing.

3. Add 0.2 ml of lysis solution. Also add 0.2 ml neutralized phenol/chloroform/isoamyl alcohol (25:24:1) and 0.3 g acid-washed glass beads.

4. Vortex at 16 000 *g* for 2 min.

5. Spin at 16 000 *g* for 5 min. Transfer the supernatant to a clean tube and ethanol precipitate the DNA. Wash with 70% ethanol and dry under vacuum.

6. Resuspend the pellet in a small volume of TE (5 μl). Use 1 μl to transform *E. coli* electrocompetent cells. This procedure can also be used to recover DNA directly from yeast colonies. Use a toothpick to transfer cells from a colony directly to lysis buffer. Proceed as described above. The yields of DNA will be lower.

4.6 Use of selection strategies to enrich for positives

Selection for reporter gene expression rather than screening for the expression of a *GAL1-lacZ* reporter gene allows one to isolate positives more rapidly, and makes it easier to search large numbers of transformants for interacting proteins. Selectable reporter genes encode products required for growth on media lacking the appropriate nutrients. Strains Y153 and YPB2 carry the *HIS3* gene under the control of the *GAL1,10* promoter. When transcription of this gene is activated, these strains can grow on media lacking histidine. In practice, however, HIS3 is produced at a low constitutive level in Y153 and YPB2 such that these strains grow on media lacking histidine. Incorporation of 3-aminotriazole into this media at a concentration of 20–40 mM inhibits this basal level of HIS3 and keeps the strain from growing unless *GAL1-HIS3* expression is activated. To screen for interacting proteins, cotransform either strain with plasmids encoding both a test protein hybrid and an activation domain library and plate on media lacking histidine and containing 3-aminotriazole. These strains also contain either a *GAL1-lacZ* or a (GAL 17mer)-*lacZ* reporter gene so that His$^+$ colonies can be assayed for β-galactosidase activity.

4.7 Artefacts and troubleshooting

A variety of false positives are found when using the two-hybrid system. These plasmids cause the reporter strain to turn blue in a β-galactosidase assay, but do not encode an interacting protein. In some cases, these plasmids alone will activate transcription of the reporter gene, but in other cases, they appear to require the DNA-binding domain hybrid for their activity. Methods for differentiating the fasle-positives from the desired true-positives are discussed below.

Screening yeast libraries prepared from a *GAL4*$^+$ strain will result in isolating plasmids encoding functional GAL4. These can easily be screened out by one of several methods.

(a) Because the activity of functional GAL4 clones is independent of the DNA-binding domain hybrid, a loss of the plasmid encoding this hybrid (by growing on nonselective media) allows one to test for plasmid dependence of the blue colour.

(b) The presence of the *GAL4* sequence can be assayed directly by Southern blot or PCR.

(c) A time-consuming method is to isolate the plasmid from each positive and to retransform the reporter strain and check for plasmid dependence.

Searches performed using a DNA-binding domain with a different binding specificity (such as LexA) will avoid this type of positive.

More commonly seen, but harder to identify, false-positives have been identified in searches of both yeast and mammalian libraries. These are clones that appear to be dependent on the target protein hybrid for activity, but in fact activate with a variety of DNA-binding domain hybrids that are unrelated to the target protein. It is likely that these clones encode activation domain hybrids that do not bind to the test protein hybrid but instead bind to or regulate promoter sequences flanking the GAL4 binding sites. The presence of a DNA-binding domain hybrid stimulates the activation of the reporter gene with this type of false-positive. Testing candidates against a variety of DNA-binding domain hybrids can identify clones that show activity specifically in the presence of the appropriate target protein hybrid. Additionally, testing candidates in strains with two reporter genes under the control of dissimilar promoters can differentiate false-positives from the desired clones. Hybrids encoding proteins that bind to the target protein hybrid will activate the reporter gene if it contains upstream binding sites for GAL4 regardless of the sequence of the flanking DNA. False-positives, however, appear to require specific sequences for their activity such that they may be unable to activate transcription of a reporter gene under the control of a promoter unrelated to the one used in the original search. Strains Y153 and YPB2 employ two different reporter genes (diagrammed in *Figure 4*) under the control of dissimilar promoters. Requiring that strains carrying the candidates and the target protein hybrid be both His$^+$ and produce β-galactosidase eliminates many false-positives. Putative true-positives should be tested further to determine that their activity is specific for the appropriate test protein hybrid. In any case, interactions identified by the two-hybrid approach are best verified by an independent methodology. For example, a gene identified from the library can be transferred to an expression vector, and the encoded protein purified and fixed to a chromatographic column. This column is then used to fractionate a cellular extract. Bound proteins are eluted, and the presence of the target protein in this fraction can often be detected by an immunoassay.

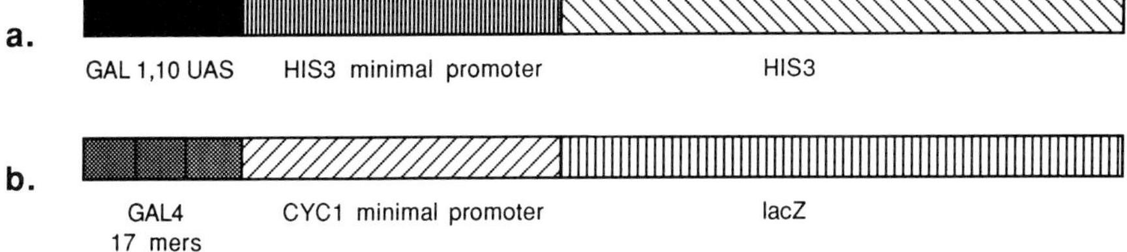

Figure 4. Diagram of the reporter genes in yeast strain YPB2 (not drawn to scale). (a) One reporter gene consists of the ~150 bp *GAL1, 10* UAS, which contains four GAL4 binding sites, fused to the *HIS3* minimal promoter and the yeast *HIS3* gene (32). (b) The other reporter gene consists of three consensus GAL4 binding sites fused to the *CYC1* minimal promoter and bacterial *lacZ* (33). Hybrids encoding proteins that interact with the target protein, which is fused to the DNA-binding domain, will activate transcription of both reporter genes, whereas false-positives will usually activate transcription of only one of the reporter genes.

5. Other applications

In addition to its use in identifying interactions between known proteins and proteins encoded by an activation domain library and a test protein, the two-hybrid system can be used to map the domains and specific residues involved in protein interactions and might be used to construct a protein linkage map which identifies all the detectable protein interactions of a test organism.

5.1 Mapping interacting domains

Producing fusions carrying different fragments of the gene encoding an interacting protein may identify a minimal protein domain that interacts with a test protein. The truncated proteins should be produced as activation domain hybrids to avoid the possibility of revealing formerly hidden activation domains in hybrids with the DNA-binding domain. The lack of a signal when assaying protein fragments does not mean that the domain being tested does not interact with the test protein. For example, the fragment may not form a stable or correctly folded hybrid. Thus only a positive signal with a truncated hybrid can identify the interacting domain.

5.2 Mutational analysis

Specific residues involved in protein–protein interactions can be identified by testing mutant protein hybrids in the two-hybrid system. Site-specific mutations of codons for residues likely to be involved in the interaction or random mutations can be introduced into one of the two test protein hybrids. The effect of these mutations can be assayed by introducing plasmids encoding the mutant hybrid and a hybrid of its wild-type protein partner into a reporter strain. The strength of the protein–protein interaction appears to correspond to the level of β-galactosidase activity. Strains carrying loss-of-binding mutants should be tested by Western blot to ensure that a sufficient level of full-length hybrid is still being produced. Theoretically, mutants that give a stronger signal and, therefore, bind more strongly can also be isolated. It should also be possible to use a mutant version of one test protein and screen mutant pools of the other test protein to identify compensating mutations that restore a protein–protein interaction.

5.3 Protein linkage maps

The two-hybrid system might be used to find large numbers of interacting protein pairs by screening random libraries of both DNA-binding domain hybrids and activation domain hybrids. Both libraries are simultaneously introduced into a reporter strain and a selection for transformants expressing reporter genes identifies interacting protein pairs. The use of a conditional promoter for the gene encoding the activation domain hybrid can rapidly eliminate positives due solely to DNA-binding domain hybrids. A partial DNA sequence for each protein in a pair is obtained and serves to 'tag' the interaction in a database. The tabulation of a large number of interactions is used to construct a 'protein linkage map' which describes the network(s) of detectable interactions between proteins in the cell. This information would identify new protein–protein interactions and would help place novel proteins in some cellular pathway. A protein linkage map, in conjunction with complete genomic or cDNA sequences, would be valuable in translating the one-dimensional DNA information into more of a three-dimensional picture of the cell.

6. Conclusions

The two-hybrid system can be used to identify interactions between known proteins and to identify, from a library, proteins that interact with a protein of interest. We have presented methods to use the vectors and strains with which we are most familiar, although other vectors and strains are being generated. In the near future, it is likely that a variety of DNA-binding and activation domain vectors, activation domain libraries, and selectable strains will be available that will make it easier to use the two-hybrid system.

Acknowledgements

We thank Joe Lipsick for comments on the manuscript, Steve Elledge and Mark Johnston for many helpful discussions, and Carol Carter and Steve Elledge for sharing unpublished data. This work was supported by NIH grants GM28220 (to R.S.) and CA54699 (to S.F.).

References

1. Young, R. A. and Davis, R. (1983). *Proc. Natl Acad. Sci. USA*, **80**, 3187.
2. Sikela, J. M. and Hahn, W. E. (1987). *Proc. Natl Acad. Sci. USA*, **84**, 3038.
3. McGregor, P. F., Abate, C., and Curran, T. (1990). *Oncogene*, **5**, 451.
4. Blackwood, E. M. and Eisenmann, R. N. (1991). *Science*, **251**, 1211.
5. Skolnik, E. Y., Margolis, B., Mohammadi, M., Lowenstein, E., Fischer, R., Drepps, A., Ullrich, A., and Schlessinger, J. (1991). *Cell*, **65**, 83.
6. Defeo-Jones, D., Huang, P. S., Jones, R. E., Haskell, K. M., Vuocolo, G. A., Hanobik, M. G., Huber, H. E., and Oliff, A. (1991). *Nature*, **352**, 251.
7. Keegan, L., Gill, G., and Ptashne, M. (1986). *Science*, **231**, 699.
8. Hope, I. A. and Struhl, K. (1986). *Cell*, **46**, 885.
9. Ma, J. and Ptashne, M. (1987). *Cell*, **48**, 847.
10. McKnight, J. L. C., Kristie, T. M., and Roizman, B. (1987). *Proc. Natl Acad. Sci. USA*, **84**, 7061.
11. Ma, J. and Ptashne, M. (1988). *Cell*, **55**, 443.
12. Brent, R. and Ptashne, M. (1985). *Cell*, **43**, 729.
13. Fields, S. and Song, O.-K. (1989). *Nature*, **340**, 245.
14. Chien, C.-T., Bartel, P. L., Sternglanz, R., and Fields, S. (1991). *Proc. Natl Acad. Sci. USA*, **88**, 9578.
15. Dalton, S. and Treisman, R. (1992). *Cell*, **68**, 597.
16. Kato, G. J., Lee, W. M. F., Chen, L., and Dang, C. V. (1992). *Genes and Dev.*, **6**, 81.
17. Vasavada, H. A., Ganguly, S., Germino, F. J., Wang, Z. X., and Weissman, S. (1991). *Proc. Natl Acad. Sci. USA*, **88**, 10 686.
18. Goff, S. A., Cone, K. C., and Chandler, V. L. (1992). *Genes and Dev.*, **6**, 864.
19. Silver, P. A., Keegan, L. P., and Ptashne, M. (1984). *Proc. Natl Acad. Sci. USA*, **81**, 5951.
20. Ma, J. and Ptashne, M. (1987). *Cell*, **51**, 113.
21. Gill, G. and Ptashne, M. (1987). *Cell*, **51**, 121.
22. Sambrook, J., Fritsch, E. F., and Maniatis, T. (1989). *Molecular cloning: a laboratory manual*. Cold Spring Harbor Lab., Cold Spring Harbor, NY.
23. Guthrie, C. and Fink, G. R. (ed.) (1991). *Methods in enzymology* Vol. 194. Academic Press, San Diego, CA.
24. Ito, H., Fukada, Y., Murata, K., and Kimura, A. (1983). *J. Bacteriol.*, **153**, 163.
25. Schiestl, R. H. and Gietz, R. D. (1989). *Curr. Gen.*, **16**, 339.
26. Hill, J., Donald, K. A., and Griffiths, D. E. (1991). *Nucl. Acids Res.*, **19**, 5791.
27. Breeden, L. and Nasmyth, K. (1985). *Cold Spring Symposia on Quantitative Biology*, Vol. 50, p. 643 (Cold Spring Habor Lab., Cold Spring Harbor, NY).
28. Miller, J. H. (1972). *Experiments in molecular genetics*. Cold Spring Harbor Lab., Cold Spring Harbor, NY.
29. Hardy, C. F. J., Sussel, L., and Shore, D. (1992). *Genes and Dev.*, **6**, 801.
30. Yang, X., Hubbard, E., and Carlson, M. (1992). *Science*, **257**, 680.
31. Hoffman, C. S. and Winston, F. (1987). *Gene*, **57**, 267.
32. Flick, J. S. and Johnston, M. (1990). *Mol. Cell. Biol.*, **10**, 4757.
33. Chasman, D. I., Leatherwood, J., Carey, M., Ptashne, M., and Kornberg, R. D. (1989). *Mol. Cell. Biol.*, **9**, 4746.
34. Luban, J., Alin, K. B., Bossolt, K. L., Humaran, T., and Goff, S. P. (1992). *J. Virol.*, **66**, 5157.

8

Purifying and assaying mesoderm-inducing factors from vertebrate embryos

JIM C. SMITH

1. Introduction

The first cell–cell interaction in amphibian development is mesoderm induction, when a signal from the vegetal hemisphere of the embryo induces overlying equatorial cells to form mesoderm rather than ectoderm (see *Figure 1*). The last five years have seen a great increase in our understanding of mesoderm induction, largely due to the discovery of several 'mesoderm-inducing factors'— molecules which induce prospective ectodermal tissue of the amphibian embryo to differentiate as a variety of mesodermal cell types such as muscle, notochord, mesenchyme, and mesothelium. Most of these inducing factors are members of the transforming growth factor type β (TGFβ) or fibroblast growth factor (FGF) families (1), and one of the most potent, activin, has been shown to induce expression of mesoderm-specific genes, or of mesodermal cell behaviour, in mouse (2), bird (3), and fish (4), as well as amphibian embryos. This indicates that mesoderm induction occurs in all vertebrates and the process is thus of considerable interest and importance.

The discovery of mesoderm-inducing factors and, more recently, of factors which alter mesodermal pattern, depended on the design of suitable assays, based on knowledge derived from experimental embryological experiments. This chapter describes two of these assays, and discusses two 'case histories' in which such agents have been discovered and purified.

2. *Xenopus* embryos

The most widely-used amphibian species is the South African clawed frog *Xenopus laevis*, an animal well-suited to laboratory conditions because these are invariably better than those encountered by the animals in the wild (5). For relatively small-scale work involving *Xenopus* embryos it is necessary to

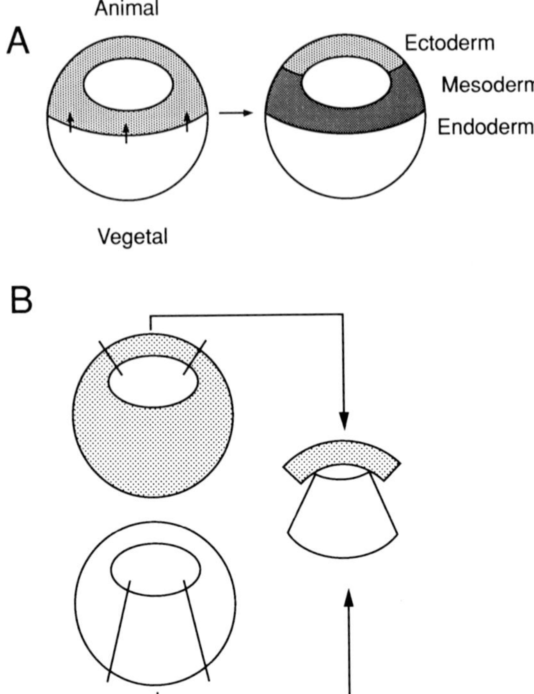

Figure 1. Demonstration of mesoderm induction. (A) A signal derived from the vegetal hemisphere of the blastula-staged embryo induces overlying tissue to form mesoderm. In the absence of such a signal, the cells would form ectoderm. (B) Mesoderm induction can be demonstrated by juxtaposing tissue from the animal pole of the embryo (which normally forms epidermis) with cells from the vegetal hemisphere. The animal pole cells now form mesodermal cell types such as muscle, notochord, and blood. It is possible to label the animal pole cells with a cell lineage marker to deduce the provenance of the mesodermal tissue.

establish a colony of at least 100–200 females and perhaps 50 males to provide testes for artificial fertilization.

2.1 A *Xenopus* colony

Different laboratories will set up their colonies in different ways, depending mostly on space and financial considerations. At one extreme Wu and Gerhart (5) describe tanks 4.2 m long and 1.2 m wide which can accommodate 1500 full-grown adults. The total volume of water is about 800 litres and a portion of this is changed daily. Although these tanks are popular with the frogs, they do not meet with the approval of the animal inspectors because they 'look like a natural pond' (5), and Gerhart's laboratory has been forced to resort to tanks 1.8 m long, 0.75 m wide and 0.6 m deep, through which water passes continuously. Such a system is certainly highly desirable, but for most laboratories, and especially those where *Xenopus* is not the main experimental material, smaller tanks will suffice with the water being changed manually every two or three days. One frog per 3 litres of water is recommended by Wu and Gerhart (5) for optimal growth, but space considerations may dictate higher population densities.

Water temperature should be maintained between 18°C and 22°C. Some workers keep their *Xenopus* on a natural light/dark cycle, whereas others prefer 12 h light/12 h dark. Both have their advantages and disadvantages: the former allows for rapid growth of *Xenopus* in spring, but eggs are generally of poorer quality in late summer and autumn; the latter gives eggs of uniform (albeit perhaps lower) quality throughout the year, but there is no increased growth in the spring.

Xenopus are fairly tolerant of water quality, but in some areas (such as London) the tap water is treated with chlorine, and this is somewhat toxic, especially to young animals. To overcome this the water can be allowed to stand for a day or so to allow the chlorine to evaporate, it can be treated with commercial aquarium products designed for this purpose, or it can be passed through charcoal filters and ion exchange columns.

Similarly, *Xenopus* are not fussy eaters. Raw meat is convenient, because it can be cut and frozen in handy bite-sized chunks. Dry food such as Nasco 'frog brittle' is also popular with some workers, while others like to feed live food, such as meal worms or tubifex. Frogs should be fed two or three times a week, and excess food should be cleared from the tank an hour or so after feeding.

This account of *Xenopus* maintenance is necessarily brief, but should provide enough information for those wishing to keep relatively small numbers of animals. More ambitious workers are referred to Wu and Gerhart (5), who also include information about diseases to which *Xenopus* is susceptible and about rearing frogs.

2.2 Obtaining eggs, fertilizing and culturing them, and staging the embryos

2.2.1 Obtaining eggs

It is usually more convenient to obtain *Xenopus* embryos by artificial fertilization than through 'natural matings', because in this way the embryos develop synchronously. Two or three days before injection of human chorionic gonadotrophin (HCG; Sigma) females (which are larger than males and have prominent cloacal flaps) can be 'primed' by injection of 30 IU pregnant mare serum gonadotrophin (PMSG; Sigma). This step is not essential, but does increase the number of ovulated eggs. HCG injection (500–1000 IU) should be carried out 10–12 h before the desired time of ovulation. Injection volumes should be about 0.5 ml, and injections should be made into the dorsal lymph sac of the animal using a 23–25 gauge needle. The simplest procedure (if one is right-handed) is to immobilize the frog with the left hand and to use the right hand to insert the injection needle just beneath the skin on the dorsal surface of the frog's right hind leg. Then, slide the tip beneath the row of 'stitches' of the lateral line organs towards the dorsal midline. The injection should be delivered here, and as the needle is withdrawn the skin can be gently massaged to prevent leakage of PMSG or HCG. If done properly there should be no bleeding and the animal should not struggle or appear to suffer any discomfort. Note that injection of *Xenopus* in this way is regarded as a 'procedure' under the UK Animals (Scientific Procedures) Act of 1986, and requires a Home Office licence.

2.2.2 Artificial fertilizations

To obtain testes, male *Xenopus* should be killed by decapitation and rapid rostral and caudal pithing. Make an incision in the lower abdomen and remove the testes, which are pale, slightly curved organs about 1 cm long positioned either side of the spine. Testes can be stored at 4°C in 60% Leibovitz L-15 medium for 3–4 days.

Expel eggs from a treated female by gentle peristalsis of her ventrolateral surface. The eggs can be collected on a small sheet of aluminium foil and then transferred to a Petri dish. Rinse the eggs briefly with distilled water, and then remove as much liquid as possible using a Pasteur pipette. Rub the eggs lightly with a piece of dissected testis, teased open to allow egress of sperm. After 5 min cover the fertilized eggs with water or 10% Normal Amphibian Medium (NAM: 110 mM NaCl; 2 mM KCl; 1 mM $Ca(NO_3)_2$; 1 mM $MgSO_4$; 0.1 mM Na_2EDTA; 2 mM sodium phosphate, pH 7.5; 1 mM $NaHCO_3$; 50 µg/ml gentamycin). After about 15 min, the eggs should rotate so that the heavily-pigmented 'animal hemisphere' is uppermost, and cleavage should begin about 90 min after fertilization.

If the fertilization is unsuccessful, the fault usually lies with the eggs and it may be necessary to try another female. Sperm can be tested for viability by making a suspension in a small volume of water and observing them under the phase-contrast microscope; a significant number should be motile.

2.2.3 Dejellying embryos

The jelly surrounding fertilized eggs and embryos should be removed by incubation for about 5 min in 2% cysteine hydrochloride (Sigma) adjusted to pH 7.8–8.1 with NaOH. The embryos should then be thoroughly rinsed in several changes of 10% NAM. They can be moved from dish to dish using a blunt Pasteur pipette.

2.2.4 Culture of embryos

Embryos should be cultured until the desired stage in 10% NAM at temperatures between 14°C and 23°C. They should be staged according to the Normal

Table of Nieuwkoop and Faber (6). If this proves difficult to obtain, photographs of the different stages are shown in reference 7. The rate of *Xenopus* development depends on temperature, which can be varied according to the investigator's requirements. It takes about 5 h at 22°C for embryos to reach the midblastula stage (stage 8), which is when the animal cap assay is usually performed.

3. The animal cap assay

The discovery of the mesoderm-inducing activities of activin and FGF depended on the so-called animal cap assay, which originated with the work of Nieuwkoop (8), who showed that presumptive ectodermal cells from the animal hemisphere of the blastula-staged embryo could be induced to form mesoderm by a signal from the vegetal hemisphere.

3.1 Dissection instruments

Dissection of *Xenopus* embryos requires the following pieces of equipment.

3.1.1 Watchmaker's forceps

No. 5 Dumont watchmaker's forceps are used to remove the vitelline membrane from embryos. Keller (9) recommends using an 'Arkansas' sharpening stone to make one pair blunt, but with ends that meet precisely, and the other with sharply tapering points, that also meet exactly.

3.1.2 Tungsten needles

About 5 cm of tungsten wire (0.5 mm diameter, 99.95% purity: Goodfellow) should be mounted in a length of capillary tubing (about 7 cm). Sharpen the needle by electrolysis in 1 M NaOH using a potential difference of 9–12 V alternating current. A microscope transformer is a useful power source. Place a small piece of modelling clay on the tip of the needle and immerse this, and about 2 mm of naked tungsten wire, in the NaOH; when the clay falls off the needle is acceptably sharp.

3.2 Animal cap dissections

Animal caps to be used in assays of mesoderm-inducing activity should be dissected from embryos at the midblastula stage (stage 8 of Nieuwkoop and Faber). Transfer the embryos to an agarose-coated Petri dish (1% agarose in water) in 75% NAM. Remove the vitelline membranes from a group of embryos using sharpened forceps, and dissect animal pole regions from the centre of the pigmented hemisphere of the embryos using a tungsten needle. The easiest way of doing this is to support the embryo with a pair of forceps and to make a gentle 'picking' movement with the needle. Alternatively, it is possible to use sharpened forceps as a pair of 'scissors'. The 'standard' animal cap is a disc of tissue from the centre of the pigmented hemisphere subtending a solid angle of about 60° (10), although it is possible to take slightly larger pieces without autonomous mesoderm differentiation, and for the sake of speed most people find it easier to cut a square piece of tissue. With practice it becomes possible to dissect about 100–200 animal caps per hour. Use a Pasteur pipette to transfer animal caps to test solutions, making sure that the tissue does not meet the meniscus of any solution. Caps should be cultured at 18–22°C.

3.3 Scoring the result

The result of the animal cap assay can be assessed in several ways. In increasing order of complexity these are described below.

3.3.1 Gastrulation-like movements

The simplest method takes advantage of the dramatic gastrulation-like movements that are induced by mesoderm-inducing factors such as activin (11). Normally, isolated animal pole regions 'round up' after being dissected from the embryo and form a sphere; they eventually differentiate as 'atypical epidermis'. The first response of those exposed to activin, however, is to gastrulate, and this manifests itself through a change in shape, involving a dramatic elongation and constriction of the tissue. This change in shape is detectable within a few hours and is clearly visible after overnight incubation. It represents, therefore, a rapid, simple, and reliable indicator of induction (11) which was of great use in the purification of the mesoderm-inducing activity secreted by XTC cells (12).

3.3.2 Later morphology

The gastrulation-like movements induced by activin are much more dramatic than those induced by FGF, while other mesoderm-inducing agents, such as bone morphogenetic protein 4 (13), appear not to induce gastrulation movements at all. For factors such as these, it is simplest to allow the induced tissue to differentiate for 3–4 days at 18°C. Uninduced animal caps remain as opaque spheres of atypical epidermis, while induced caps form more normal, translucent epidermis with the internal structure clearly visible within. Activin, and others agents which induce dorsal mesoderm, can elicit the formation of structures which have been termed 'embryoids', due to their resemblance to intact, whole embryos (14). Agents which induce ventral mesoderm, such as FGF, induce balloon-shaped structures with a layer of smooth muscle beneath the epidermis. These are most clearly illustrated in histological preparations (see below).

3.3.3 Histology

Identification of particular cell types in induced animal pole regions can best be achieved by histological analysis of specimens after 3–4 days culture.

Explants should be fixed for 24–48 h in 10% formalin, 2% glacial acetic acid, 50% ethanol, and 38% NAM. After fixation, specimens should be rinsed in 70% PBS for 30 min, then in 50% ethanol for 30 min, after which they can, if necessary, be stored in 70% ethanol.

The specimens can then be embedded in paraffin wax via butanol according to *Protocol 1*.

Protocol 1. Embedding specimens for histology

Immerse specimens in the following solutions for the times and temperatures indicated:

1. 80% ethanol   30 min, room temperature
2. 60% ethanol, 20% butanol, 20% H_2O 30 min, room temperature
3. 45% ethanol, 45% butanol, 10% H_2O 30 min, room temperature
4. 25% ethanol, 75% butanol   30 min, room temperature
5. butanol   1 h, room temperature
6. 50% butanol, 50% paraffin wax   20 min, 60°C
7. paraffin wax   2 changes, each for 1 h, 60°C
8. Embed in fresh paraffin wax; allow to set

Cut sections at 7–10 μm and mount on 'subbed' slides. These should be prepared by dipping them in a solution of 0.25% gelatin and 0.025% chromic potassium sulfate and allowing them to dry in a dust-free atmosphere.

Stain sections by the Feulgen/Light Green/Orange G technique (*Protocol 2*).

Protocol 2. Feulgen/Light Green/Orange G staining technique

Reagents

Schiff's reagent:

- Add 5 g pararosaniline HCl to 150 ml 10% HCl.
- To this add 5 g potassium metabisulfite in 850 ml H_2O. Stir overnight in the dark.
- Add 2.5–3 g activated charcoal. Stir for at least 1 h.
- Filter. If the solution is slightly pink, add more activated charcoal and repeat step 3.
- Store at 4°C.

Protocol 2. *Continued*

Sulfite rinse:

- For enough reagent to fill one slide rack, mix 810 ml H_2O, 45 ml 10% HCl, and 45 ml 10% potassium metabisulfite.

Orange G

- 1% Orange G in 2% phosphotungstic acid.

Light Green

- 0.25% Light Green in 95% ethanol.

Method

1. Transfer sections to distilled water by 1 min immersions in xylene (twice), 100% ethanol (twice), 95% ethanol, 80% ethanol, 70% ethanol, 50% ethanol, and water (twice).
2. Hydrolyse in 5 M HCl for 1 h.
3. Rinse twice in distilled water.
4. Incubate in Schiff's reagent 1 h (longer if not fresh).
5. Give three sulfite rinses, 3 min each.
6. Leave in running tap water at least 0.5 h.
7. Rinse three times in distilled water.
8. Counterstain in Orange G, 1–1.5 min.
9. Wash well in three changes of distilled water.
10. Stain with Light Green, 3–4 min.
11. Rinse in water; differentiate if necessary in 70% alcohol until nuclei are clearly visible.
12. Dehydrate in 95% alcohol, 30 min.
13. Dehydrate twice in absolute alcohol, 5 min each.
14. Clear twice in histoclear or xylene, 10 min each.
15. Mount in XAM.

Cell types can be identified according to the following criteria (15, 16) and examples of different cell types are illustrated in *Figure 2*:

Notochord: cells with large internal vacuoles, stellate nuclei with prominent nucleoli

Muscle: elongated cells with longitudinal and transverse striations with large lipid-filled vacuoles

Pronephros: a single-cell thick epithelium with cuboidal cells; these also contain a large, clear vacuole

Mesenchyme: a loose network of fibroblast-like cells

Mesothelium: a thin sheet of cells usually located just inside normal epidermis; may comprise smooth muscle

Blood: Oval cells with a clear cytoplasm and condensed nuclei lacking nucleoli; not all such cells express globin and may, therefore, represent immature erythrocytes

One useful way to classify the effects of inducing factors is to distinguish between *dorsal* mesoderm, which is usually defined as that containing notochord, *intermediate* mesoderm, which is dominated by muscle, and *ventral* mesoderm which contains little muscle but abundant mesenchyme and mesothelium.

A more definitive identification of cell types can be achieved by the use of specific antibodies as described below.

3.3.4 Immunocytochemistry

Many antibodies are available to distinguish between different cell types in *Xenopus* embryos. Those most frequently used to identify mesodermal cell

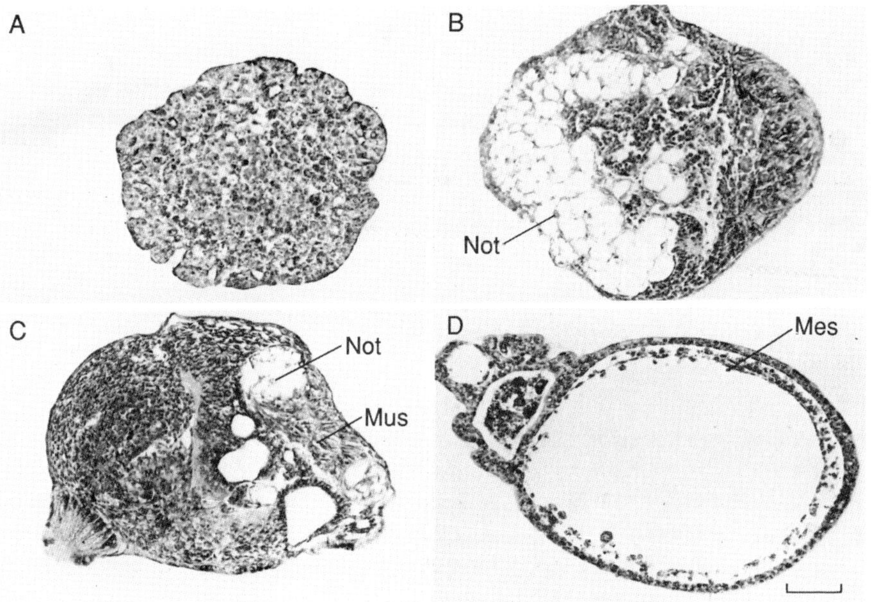

Figure 2. Cell types formed in response to activin and bFGF. (A) Animal pole regions cultured in the absence of mesoderm-inducing factors form 'atypical epidermis'. (B, C) Animal pole regions treated with activin. Note notochord (not) and muscle (mus). (D) Animal pole region treated with FGF. Note the formation of normal epidermis (epi), with a layer of mesothelium just beneath (mes). Scale bar in (D) applies to all figures, and is 100 μm.

types include MZ15 (17) and Tor 70 (18) to identify notochord, 12/101 (19) and an anti-*Xenopus* myosin heavy chain antiserum to identify muscle (20), and an anti-*Xenopus* globin antiserum to recognize blood (21). There are also antibodies which recognize epidermis (22) and neural tissue (23). These antibodies can be used on sectioned material or on intact explants, as 'whole-mounts'. The advantage of cutting sections is that different antibodies can be applied to adjacent sections, and it is possible to get a clearer idea of the overall structure of the explant; the advantage of 'whole-mounts' is that it is not necessary to cut sections. Methods for both approaches are given here.

i. Sectioned specimens

Different antibodies may require different fixation protocols, of which two are:

(a) 2% trichloroacetic acid in water

(b) 4% paraformaldehyde in 70% phosphate-buffered saline

With both fixatives, treatment for 1 h at room temperature, or overnight at 4°C, is sufficient. The harsh solvents used in conventional histology (see above) frequently disrupt antigenic determinants; using frozen sections overcomes this problem. One technique that works well with embryonic tissue is given in *Protocol 3* (reference 24).

Protocol 3. Embedding specimens in acrylamide

Reagents

- AEM: 8.4 g acrylamide, 13.4 mg bisacrylamide, 100 µl TEMED, 70% phosphate-buffered saline (PBS) to 100 ml

Method

1. Wash specimens thoroughly in PBS.

2. Incubate specimens in acrylamide embedding medium (AEM) for 5–18 h at 4°C.

3. Pre-polymerize a layer of acrylamide on the bottoms of embedding moulds by adding 50 µl 10% ammonium persulfate (APS) to 10 ml AEM and placing enough in each mould to give a layer of about 2 mm thick. Allow this to polymerize by putting the moulds in a vacuum desiccator attached to a water pump.

4. When the cushion has polymerized fill the moulds completely with AEM plus ammonium persulfate (50 µl APS per 10 ml of AEM as above), and add specimens. Cover the moulds with a layer of Parafilm to exclude air, and allow to polymerize overnight at 4°C.

5. Freeze the blocks on dry ice/ethanol. They may be stored in a humidified box at $-20\,°C$.

6. Trim the block to shape and cut 10 μm sections on a cryostat at -20 to $-25\,°C$. Mount sections on subbed slides (see Section 3.3.3).

Sections may be stored in a humidified box at $4\,°C$ for at least a week before use. Process them for immunofluorescence according to *Protocol 4*.

Protocol 4. Indirect immunofluorescence

1. Rinse slides in two changes of PBS.

2. 'Block' nonspecific binding of the second antibody by pre-incubating slides for 20 min with 4% bovine serum albumin (BSA) in PBS plus 10% normal serum. Use serum derived from the species in which the second (fluorescent) antibody was raised. Keep the slides horizontal in a humidified atmosphere and apply the blocking solution in a 'puddle' on the slides.

3. Drain excess fluid on to a paper tissue.

4. Add primary antibody diluted in 1% BSA in PBS, as described for the blocking solution (step 2). The concentration of primary antibody depends on its titre, and may be anything from 10% to 0.1%; this should be determined empirically. Incubate 1 h at room temperature or overnight at $4\,°C$.

5. Drain slides and wash three times in PBS, 15 min each.

6. Add the second antibody diluted in 1% BSA in PBS, as described for the blocking solution (step 2). If the primary antibody was the muscle marker 12/101, for example, which is a mouse IgG, this second antibody might be fluorescein isothiocyanate-conjugated rabbit-anti mouse IgG. Incubate 1 h at room temperature in the dark. The second antibody should be centrifuged before use and used at a concentration of about 2%, although as with the primary antibody it is best to determine the optimum concentration empirically.

7. Drain slides and wash for 15 min in PBS.

8. Counterstain by incubating sections in the following solution for 5 min:
   - 2 μl DAPI (1 mg/ml; DAPI, 4′,6-diamidino-2-phenylindol-dihydrochloride)
   - 40 μl Eriochrome Black (1%; Eriochrome Black is also known as Solochrome Black and Mordant Black III)
   - 10 ml PBS

Protocol 4. *Continued*

This step may be omitted, but the DAPI provides a useful way of viewing the nuclei, and the Eriochrome Black, which appears orange/red under fluorescein illumination, effectively counteracts the autofluorescence of the yolk platelets of *Xenopus*.

9. Drain slides and wash twice in PBS, 15 min each.

10. Mount slides in Citifluor glycerol/PBS solution (Citifluor Ltd) or Gelvatol.

Mounting slides in Gelvatol (25)

1. Add 2.4 g Gelvatol 20–30 (Monsanto) to 6 g glycerol, and mix.

2. Add 6 ml water and leave for several hours at room temperature.

3. Add 12 ml 0.2 M Tris (pH 8.5) and heat to 50°C for 10 min with occasional mixing until the Gelvatol dissolves.

4. Centrifuge at 5000 *g* for 15 min.

5. Add 1,4-diazobicyclo-[2.2.2]-octane (DABCO) to 2.5% to reduce quenching of fluorescence.

6. Aliquot in air-tight containers and store indefinitely at −20°C.

---

ii. Whole-mount immunocytochemistry

Two fixation protocols have been recommended for whole-mount immunocytochemistry:

- 80% methanol/20% DMSO, −20°C overnight
- MEMFA, 1–2 h at room temperature

MEMFA is 0.1 M MOPS (pH 7.4), 2 mM EGTA, 1 mM $MgSO_4$, 3.7% formaldehyde. Other fixation protocols, of course, may also be appropriate. After fixation transfer specimens to methanol and store at −20°C. Process is described in *Protocol 5*.

Protocol 5. Whole-mount immunocytochemistry (based on reference 26)

1. Rinse specimens in 100% methanol, and, if necessary, bleach them in 70% methanol/10% hydrogen peroxide for either 48 h at −20°C or about 24 h at room temperature in the light. Rinse and store specimens in 100% methanol at −20°C until required. If albino embryos are used it is, of course, unnecessary to bleach them.

The following steps should be carried out with gentle agitation.

2. Rehydrate embryos three times in PBS, 15 min each.

3. 'Block' nonspecific binding of the second antibody by incubating specimens for 1 h in 1% milk powder in PBS + 0.05% Tween 20.

4. Add primary antibody to a final concentration determined empirically to give the best signal-to-noise ratio.

5. Incubate overnight at 4°C.

6. Wash six times in PBS/Tween at room temperature, 1 h each time.

7. Incubate overnight at 4°C in PBS/Tween plus approximately 0.2% HRP-conjugated second antibody (the optimum concentration of this second antibody is best determined empirically). If the primary antibody was the muscle marker 12/101, this second antibody might be HRP-conjugated rabbit-anti mouse IgG.

8. Repeat step 6.

9. Equilibrate for 30 min in 0.5 mg/ml diaminobenzidine (DAB) in PBS (**caution**: DAB is carcinogenic).

10. Add hydrogen peroxide to 0.02% and watch while the colour develops.

11. Discard DAB into bleach, wash embryos in PBS/Tween, then dehydrate in methanol, twice, 10 min each.

12. Clear in 'Murray's clear': (1 vol. benzyl alcohol:2 vol. benzyl benzoate).

3.3.5 Molecular probes

In the same way that there are now many region and cell-type specific antibodies available, there are also numerous molecular markers recognizing different mesodermal cell types or different regions of the mesoderm. Some of the most widely-used are listed in *Table 1*. These markers may be used to probe Northern blots or in RNAase protections.

RNA can be extracted from explants according to *Protocol 6*.

Table 1. Molecular probes for different regions of the mesoderm

Probe	Expression domain(s)	Reference
Goosecoid	Early gastrula: dorsal lip of the blastopore	39
Brachyury	Early gastrula: all (?) presumptive mesoderm. Early neurula: posterior mesoderm and notochord.	40
pSP21 (cardiac actin)	Neurula and later; cardiac and skeletal muscle	41
Xwnt-8	Early gastrula: ventral and lateral mesoderm	42
Xhox3	Highest in posterior mesoderm of the neurula-staged embryo. Also later expressed in neural tissue	43
XlHbox6	Lateral mesoderm of the neurula-staged embryo. Also expressed in posterior neural plate.	44

Protocol 6. Extraction of RNA from embryonic tissue (see reference 27)

Reagents

- NETS: 50 mM Tris (pH 7.5), 50 mM NaCl, 10 mM EDTA (pH 8.0), 0.5% SDS
- 5 × hybridization salts: 2 M NaCl, 0.2 M Pipes (pH 6.4), 5 mM EDTA

Method

This technique is suitable for up to five embryos worth of tissue.

1. Freeze sample in a minimum volume of medium in an Eppendorf tube on dry ice and store at −70°C.
2. Add 200 µl NETS containing 200 µg/ml proteinase K (NETS/PK; add proteinase K fresh).
3. Transfer sample in NETS/PK to a 1 ml homogenizer. Homogenize.
4. Rinse original tube with 200 µl NETS/PK and add this to the homogenizer. Homogenize again.
5. Transfer to Eppendorf containing 400 µl phenol and vortex hard. Microfuge for 5 min and transfer the aqueous phase to a fresh tube containing 400 µl phenol. Vortex and spin again.
6. Transfer the aqueous phase to a fresh tube containing 400 µl chloroform. Vortex and spin.
7. Transfer the aqueous phase to a fresh tube.
8. Add 40 µl of 3 M Na acetate, pH 5.0, and 1 ml of ethanol. Vortex and freeze on dry ice for 20 min.
9. Microfuge 10 min.
10. Remove supernatant and discard. Suspend pellet in 30 µl DEPC-treated water. Add 30 µl 8 M LiCl (DEPC-treated), mix, leave at −20°C for at least 3 h.
11. Microfuge 10 min.
12. Remove supernatant, wash with 70% ethanol, air dry pellet briefly.
13. Resuspend pellet in 2 µl water followed by 4 µl 5 × hybridization salts, 14 µl formamide (check that the pH of the formamide is not less than 7).
14. Freeze at −80°C until required.

A technique for Northern blots is given in reference 27. In general, because of the limited amounts of tissue available, it is more convenient to use the sensitive RNAase protection technique on explants of *Xenopus* embryos, and a suitable method is given in *Protocol 7*.

Protocol 7. RNAase protection on tissue from *Xenopus* embryos

Suitable molecular markers are listed in *Table 1*. To prepare antisense probes for RNAase protections the cDNAs must be cloned into a vector carrying a bacteriophage SP6, T7, or T3 promoter adjacent to a multiple cloning site. Many such vectors are available. To synthesize a probe, the plasmid should be cut at a site downstream of the promoter to yield a transcript of 150–450 nucleotides. A restriction enzyme generating blunt ends or protruding 5′ termini should be used if possible; protruding 3′ termini allow the initiation of long transcripts in the wrong direction. If it is essential to use an enzyme generating a 3′ terminus, the DNA should be rendered blunt-ended with the Klenow fragment of DNA polymerase I. Purify the template DNA by extraction with phenol:chloroform and precipitate with ethanol. Redissolve in DEPC-treated water at approximately 1 μg/ml. Make a labelled probe and analyse RNA samples as shown below.

Reagents

- $10 \times$ transcription buffer: 400 mM Tris (pH 7.5), 60 mM $MgCl_2$, 20 mM spermidine HCl
- formamide dyes: 100 ml deionized formamide, 2 ml 0.5 M EDTA, 0.1% xylene cyanol, 0.1% bromophenol blue
- elution buffer: 0.5 M NH_4 acetate, 1 mM EDTA, 0.1% SDS
- digestion buffer: 10 mM Tris (pH 7.5), 5 mM EDTA, 300 mM NaCl

Method

A. *Probe preparation*

1. Set up transcription reaction at room temperature:

   - 250 mM DTT   0.5 μl
   - RNAsin   0.5 μl
   - 10 mM rNTPs (A, C, and G)   1 μl
   - $10 \times$ transcription buffer   1 μl
   - $[\alpha\text{-}^{32}P]$rUTP (50 μCi; 450 Ci/mmol)   5 μl
   - template DNA (1 mg/ml)   1 μl
   - SP6/T7/T3 RNA polymerase   5 units

   Mix gently, spin briefly, incubate 1–2 h at 37°C (T3 and T7 polymerases) or 40°C (SP6 polymerase). It is also possible to incubate at lower temperatures, even down to 4°C; this achieves a greater yield of full-length transcripts.

2. Add 1 μl RQ1 DNAase I (Promega), incubate for 30 min at 37°C.

Protocol 7. *Continued*

3. Add 30 µl formamide dyes and load on to a short 6% sequencing gel. Run the bromophenol dye marker no more than about 16 cm.

4. Cover the wet gel with Saranwrap and autoradiograph with X-ray film for 2–3 min at room temperature. Draw marks across the edge of the film on to the gel plate to allow realignment afterwards. Using the autoradiograph as a guide, excise the labelled band into a microfuge tube. Add 0.4 ml elution buffer and leave at room temperature overnight or at 55°C for 1 h.

5. Remove the eluate to a fresh tube, add 1 ml ethanol and precipitate on dry ice for 20 min. Microfuge and resuspend in 10–50 µl RNAase-free water (depending on how many samples are to be analysed). Store at −20°C and use within a day or two.

B. *Hybridization and digestion reactions*

1. Set up 80°C and 45°C water baths or dry blocks. To each sample add 1–2 µl of the probe. Mix and spin down. Incubate at 80°C for 5 min and then transfer immediately to 45°C. Hybridize for 5 h to overnight. Meanwhile, pour a 6% thin sequencing gel.

2. Add 350 µl digestion buffer containing 2 µg/ml RNAase T1 and 40 µg/ml RNAase A to each sample. Mix and spin down. Incubate at 30 or 37°C for 30 min. (Some probes yield better results if RNAase A is omitted.)

3. Add 20 µl 10% SDS, 50 µg proteinase K. Mix and spin down. Incubate at 37°C for 10 min.

4. Add 0.4 ml phenol and vortex. Spin for 5 min in microfuge. Put 5 µg tRNA in fresh tubes and transfer aqueous phases to them. Add 1 ml ethanol, vortex, and precipitate on dry ice.

5. Spin down and remove traces of supernatant. Resuspend samples in 2 µl water and then add 2 µl formamide dyes. Boil for 1.5–3 min and load immediately on to sequencing gel. Run the gel at 40 W for 2–3 h (this can be varied according to the apparatus in use), then fix in 10% acetic acid/ 10% ethanol and dry down. Expose to X-ray film.

3.4 Quantifying mesoderm-inducing activity

It is important to define a 'unit' of mesoderm-inducing activity so that the specific activities of different preparations can be compared and so that recoveries of inducing activity during purification can be assessed. One unit of mesoderm-inducing activity is defined as the minimum amount that must be present in 1 ml of medium in order for induction to occur (28). This may be determined by serial dilution of the sample concerned. Thus, if a sample of medium is active at a dilution of 1 in 1000 but not at 1 in 2000, it contains 1000 units/ml.

4. RNA injection

The animal cap assay requires that proteins to be assayed are available in a soluble and, preferably, a purified form. This is not always possible, of course, and an alternative is to microinject mRNA encoding the protein of interest into the animal hemisphere of the embryo, where it can be translated and, if appropriate, secreted. It is then possible to score the results by dissecting animal pole tissue from the embryo and allowing it to develop in isolation, in the absence of inducing factors, or by observing the effect of RNA injection on whole embryos. In experiments where animal caps are dissected from embryos, the methods of analysis are the same as those used in the animal cap assay (above). Where the phenotypes of whole embryos are being studied the simplest approach is to allow the embryo to reach tailbud stages and to observe the pattern of the body plan. A useful addition to this approach is to make use of embryos which have been made radially ventral by UV irradiation of their vegetal hemispheres before first cleavage. In this case, it is possible to identify mRNAs which have the effect of 'dorsalizing' these ventralized embryos; that is, of restoring them to normality. Techniques for synthesizing RNA for injection into *Xenopus* embryos, for UV-irradiating fertilized eggs to make them radially ventral, and for microinjection of mRNA are described below.

4.1 RNA synthesis

Although it is possible to synthesize RNA from any vector containing an SP6, T7, or T3 promoter, the most stable transcripts, and those from which translation is most efficient, are obtained from the vector pSP64T, which includes globin 5′ and 3′ flanking regions together with a poly A tail (29). It is also important to include a 5′ cap on the RNA for efficient initiation of translation. *Protocol 8* describes the preparation of RNA for injection into *Xenopus* embryos.

Protocol 8. Preparation of RNA for microinjection into *Xenopus* embryos

Reagent

- 10 × transcription buffer: 400 mM Tris (pH 7.5), 60 mM $MgCl_2$, 20 mM spermidine HCl, 50 mM NaCl.

Method

1. Set up transcription reaction at room temperature:
   - DNA (1 mg/ml)   5 μl
   - 10 × transcription buffer   5 μl

Protocol 8. *Continued*

- 0.25 M DTT   2.5 µl
- 10 mM ATP   5 µl
- 10 mM CTP   5 µl
- 10 mM UTP   5 µl
- 1 mM GTP   5 µl
- 2.5 mM GpppG   10 µl
- RNAsin   2.5 µl
- SP6 RNA polymerase   2.5 µl
- H_2O 2.5 µl

Mix gently, spin briefly, incubate for 30 min at 37°C.

2. Add 2.5 µl 10 mM GTP and incubate for an additional 1 h.

3. Add 5 µl of RQ1 DNAase I (Promega) and incubate for 30 min at 37°C.

4. Extract with phenol/chloroform and ethanol precipitate twice. Wash pellet twice with 75% ethanol to remove unincorporated nucleotides.

5. Re-dissolve RNA in DEPC-treated water, measure A_{260} (expect about 50 µg total RNA) and adjust to desired concentration (say, 1 mg/ml).

It is possible to test the reaction 30 min into the final incubation by taking 1 µl of the reaction and running it on a 1% TBE gel (RNAase free), comparing it with 100 ng of cut plasmid. There should be a strong RNA band of the appropriate size. The translatability of the RNA can also be tested by *in vitro* translation.

RNA can be stored at −80°C until use. It should be centrifuged immediately before loading into the injection pipette.

4.2 Microinjecting *Xenopus* embryos

A difficulty in describing techniques for the injection of *Xenopus* oocytes and eggs is that so many methods are in use. Rather than describe all of these I shall concentrate on the approach used in my laboratory. Descriptions of other techniques have been given by Colman (30) and more recently by Kay (31).

4.2.1 Needles

Micropipettes are prepared from capillary tube, and needles are pulled with one of two types of puller. In the cheaper method, an old electrode puller is modified so that all that remains is the heated platinum coil. The capillary tube is clamped at its top, it passes down through the coil, and a piece of 'plasticine' modelling clay (about 7 g) is attached to its end. The amount of current passed through the coil, the weight of the modelling clay, and whether the investigator is able to catch the tubing before it hits the bench, all determine the shape of the needle. The more expensive option is to use a more modern needle puller such as that made by Campden Instruments. For injecting oocytes and fertilized eggs, the needle should have an external diameter no greater than 15 μm. For injecting RNA, needles should be baked at 180°C for 3 hours to destroy any RNAase activity. They should be stored in a dust-free environment.

4.2.2 Injection system

A simple injector is the Inject+Matic, an air-driven system in which air pressure and pulse duration can be controlled so as to vary the injection volume (Micro Instruments). The Inject+Matic is also capable of applying negative pressure, so that injection samples can be sucked up into the pipette.

4.2.3 Micromanipulator

The simplest micromanipulator is the Mark I model of the Singer Instrument Company.

4.2.4 Calibrating the pipettes

It is necessary to calibrate each pipette individually, so if several samples have to be injected it is more convenient to use the same pipette and rinse it thoroughly with DEPC-treated water between each sample. The most convenient way to calibrate pipettes is to expel liquid into a dish of sterile mineral oil placed on top of a graticule, to measure the diameter of the drop, and to calculate its volume according to:

$$V = 4/3\pi r^3.$$

4.2.5 Injecting embryos

Fertilized eggs and later-staged embryos can be kept at 14°C to prolong the time available for injection. Before injection they should be transferred to 75% NAM containing 4% Ficoll. This reduces the osmotic pressure between the plasma membrane of the embryo and the vitelline membrane, and prevents leakage of cytoplasm from the embryo after injection. Injection is most easily performed (for a right-hander) by supporting the embryo with a pair of forceps in the left hand and using the right hand to control the micromanipulator to impale the embryo. After injection, each embryo can be moved to one side and the next one moved into position.

With practice, it is possible to inject several hundred embryos in an hour. They should be kept in Ficoll until at least the early blastula stage, when they can be gradually changed to 10% NAM. Such embryos can be allowed to develop until tailbud stages, or their animal pole regions can be dissected and allowed to develop in isolation in 75% NAM.

4.3 Scoring the result

The differentiation of animal pole regions derived from embryos injected with different RNA species can be followed using the techniques described above. For whole embryos, the phenotype produced by RNA injection may be rather obvious; there may arise, for example, an extra dorsal axis, as was observed after injection of int-1 RNA (32). Alternatively, there may be a loss of the dorsal axis (33, 34). In other cases, as is seen with injection of RNA encoding activin, or *Xenopus* Brachyury, there may be a conversion of presumptive ectodermal tissue into mesoderm (35). Analysis of these latter cases may require careful study of the development of the embryos, using histological, immunological, and molecular techniques.

One useful approach for agents thought to be involved in specification of the dorsal axis is to inject RNA into embryos made radially ventral by treatment with UV light shortly after fertilization. The technique is described below.

4.4 'Ventralization' of *Xenopus* embryos by UV irradiation

Dorsoventral polarity of the *Xenopus* embryo is established by a microtubule-dependent rotation of the egg cortex with respect to the cytoplasmic core. This occurs shortly after fertilization, and in normal embryos the axis of rotation is defined by the position of sperm entry, with the region of the embryo on which sperm entry occurs becoming the ventral side. Rotation can be inhibited by exposing the vegetal hemisphere of the fertilized egg to UV light and the embryos that develop are, at the extreme, radially ventral. They lack axial mesodermal structures such as notochord and muscle, and compensate by overproduction of ventral tissue such as blood (21). A more-or-less continuous spectrum of phenotypes ranging from normal to completely ventralized can be observed, and a 'dorsoanterior index' (DAI) has been devised to describe this series (36).

Treatment of *Xenopus* embryos with UV light is simple, and requires only two pieces of equipment. One consists of a 3 cm deep chamber in which the base is a quartz slide and the four walls are made of normal glass. The other is a short wavelength UV light source, either hand-held (UV Light Products Inc.) or in the form of a transilluminator of the type used for viewing ethidium bromide-stained gels.

Embryos should be dejellied using cysteine hydrochloride shortly after fertilization, and irradiation should take place within 30 min of fertilization. It is best to dejelly only those eggs that have rotated. The irradiation chamber is placed over the UV light source and is filled almost to the top with NAM. The lamp is then switched on and embryos are transferred to the chamber. During their fall through about 3 cm of 75% NAM, they orientate themselves so that they land vegetal pole-down. The time of irradiation should be determined empirically: too short, and many embryos will show only a partial phenotype; too long, and embryos will not complete cleavages in their vegetal hemispheres. As well as varying the time of irradiation, it may be convenient to adjust the distance from light source to irradiation chamber. As stated by Kao and Danilchik (37), a good starting point is a time of 0.5–2 min at a range of 1–3 cm.

After irradiation the embryos should be left undisturbed until the onset of first cleavage; if they are disturbed it is possible that artificially-induced rotation will produce some degree of 'rescue'. They can then be transferred to 75% NAM containing 4% Ficoll and injected with the RNA of interest.

5. Case-history: purification of XTC-MIF

As an example of the use of the animal cap assay in the purification of a mesoderm-inducing factor, I will briefly describe the purification of XTC-MIF, a mesoderm-inducing factor derived from the *Xenopus* XTC cell line which proved to be a *Xenopus* homologue of activin A (12). The success of this purification depended on the use of gastrulation-like movements as a marker of mesoderm formation; this simple, rapid indicator of induction permitted active column fractions to be identified and applied to the next fractionation procedure with a minimum of delay.

For preparation of large volumes of XTC-conditioned medium one can use roller bottles, Cytodex beads, or the 'Cell Factories' manufactured by Nunc. We have had success with roller bottles but particularly with Cell Factories, used according to the manufacturer's instructions. With three Cell Factories it is possible to obtain 10 litres of conditioned medium a week. We routinely concentrate this medium 10-fold using a Minitan apparatus (Millipore).

The first step in the purification of XTC-MIF involves hydrophobic inter-action chromatography. Concentrated XTC-conditioned medium is first heated to 100°C for 10 min and then rapidly cooled to room temperature. This increases the specific activity of the medium 3- to 10-fold (10). The medium is then adjusted to 1 M NaCl and applied to a column of Phenyl Sepharose CL-4B equilibrated in 20 mM Tris (pH 8.0), 1 M NaCl. The volume of sample applied to the column should be no more than five times the volume of the column. Thus, for a 5 × 30 cm column, a suitable sample volume would be approximately 3 litres. After applying the sample to the column, it should be washed with at least 1 column volume of 20 mM Tris (pH 8.0), 1 M NaCl and then with a similar volume of 20 mM Tris (pH 8.0). This step removes significant quantities of contaminating protein. Inducing activity can then be eluted with a gradient of 0–80% (w/w) ethylene glycol in 20 mM Tris (pH 8.0). The volume of the gradient should be about five times the column volume; thus for the 5 × 30 cm column described above a suitable volume would be 3 litres.

Under these conditions, inducing activity elutes at approximately 40% ethylene glycol. Active fractions from this step can be pooled and then applied to an anion-exchange column such as DEAE–Sepharose CL-4B equilibrated in 20 mM Tris (pH 8.0). It is again convenient to use a column whose volume is approximately one-fifth that of the sample to be loaded, and a suitable size is approximately 2.6 × 60 cm. After loading the sample, the column is washed with approximately one column volume of 20 mM Tris (pH 8.0), and the column is developed with a 2 litre gradient of 0–1 M NaCl; mesoderm-inducing activity elutes at approximately 0.6 M NaCl.

Active fractions from this step are acidified to pH 2.2 with trifluoroacetic acid (TFA), degassed, and pumped directly on to a 4.6 × 100 mm C8 reversed-phase high-performance liquid chromatography (HPLC) column such as the Brownlee RP300 equilibrated in 0.1% TFA. Mesoderm-inducing activity is relatively stable under acidic conditions, and all the activity is retained by the column. The column is developed at a flow rate of 1 ml/min with a steep gradient of acetonitrile (1.1%/min), and under these conditions mesoderm-inducing activity elutes at approximately 32% acetonitrile. This material can be further purified and concentrated by a second round of reversed-phase HPLC on a microbore RP300 column of dimensions 2.1 × 100 mm using an acetonitrile gradient of 0.18%/min and a flow rate of 0.2 ml/min. Material from this step is electrophoretically homogeneous and was used to determine the N-terminal amino acid sequence of XTC-MIF and thus identify it as activin A.

6. Case-history: identification of the dorsalizing activity of noggin

Activin is a powerful mesoderm-inducing factor in the animal cap assay, but when the ability of activin mRNA to dorsalize UV-irradiated ventralized embryos was tested, it was found that this could only provide partial rescue; the embryos lacked head structures. In an effort to discover factors which can provide complete rescue, Smith and Harland (38) undertook an expression cloning approach.

Poly(A)$^+$-RNA derived from embryos made hyperdorsal by treatment with LiCl at the 32-cell stage (see reference 37 for LiCl treatment of *Xenopus* embryos) was size-fractionated on a sucrose gradient, and each of the 16 fractions was tested for the ability to provide at least partial rescue of UV-irradiated embryos. One fraction displayed such activity, and a cDNA library was constructed from this RNA using the vector pGEM-5Zf(−), which places the 5′ end of the cDNA adjacent to an SP6 RNA polymerase promoter. DNA was extracted from a portion of the amplified library, and RNA synthesized from it was also capable of providing some rescue of UV-irradiated embryos.

From this point, it was a relatively simple matter to isolate single clones with dorsalizing activity by a process of sib selection, and using this approach Smith and Harland identified two cDNAs capable of providing complete rescue of UV-irradiated embryos. One of these is Xwnt-8, a *Xenopus* gene related to int-1, which McMahon and Moon had already shown to be capable of duplicating the dorsal axis of *Xenopus* (32). The other is a novel peptide the authors have named 'noggin'. Expression of noggin is restricted to the presumptive dorsal mesoderm, and it is a strong candidate for a natural dorsalizing agent in normal development.

Acknowledgements

I am grateful to Vincent Cunliffe, Raj Ladher, Helen New, and Brenda Price for their helpful comments.

References

1. Smith, J. C. (1989). *Development*, **105**, 665–77.
2. Blum, M., Gaunt, S. J., Cho, K. W. Y., Steinbeisser, H., Blumberg, B., Bittner, D., and De Robertis, E. M. (1992). *Cell*, **69**, 1097–106.
3. Cooke, J. and Wong, A. (1991). *Development*, **111**, 187–212.
4. Schülte-Merker, S., Ho, R. K., Herrmann, B. G., and Nüsslein-Volhard, C. (1992). *Development*, **116**, 1021–32.
5. Wu, M. and Gerhart, J. (1991). *Meth. Cell Biol.*, **36**, 3–18.
6. Nieuwkoop, P. D. and Faber, J. (1967). *Normal Table of* Xenopus laevis (*Daudin*). North Holland, Amsterdam.
7. Danilchik, M., Peng, H. B., and Kay, B. K. (1991). *Meth. Cell Biol.*, **36**, 679–81.
8. Nieuwkoop, P. D. (1969). *Wilhelm Roux Arch. EntwMech. Org.*, **162**, 341–73.
9. Keller, R. (1991). *Meth. Cell Biol.*, **36**, 61–113.
10. Smith, J. C. (1987). *Development*, **99**, 3–14.
11. Symes, K. and Smith, J. C. (1987). *Development*, **101**, 339–49.
12. Smith, J. C., Price, B. M. J., Van Nimmen, K., and Huylebroeck, D. (1990). *Nature*, **345**, 732–4.
13. Köster, M., Plessow, S., Clement, J. H., Lorenz, A., Tiedemann, H., and Knöchel, W. (1991). *Mech. Dev.*, **33**, 191–9.
14. Sokol, S. and Melton, D. A. (1991). *Nature*, **351**, 409–11.
15. Slack, J. M. W. and Forman, D. (1980). *J. Embryol. Exp. Morph.*, **56**, 283–99.
16. Green, J. B. A., Howes, G., Symes, K., Cooke, J., and Smith, J. C. (1990). *Development*, **108**, 173–83.
17. Smith, J. C. and Watt, F. M. (1985). *Differentiation*, **29**, 109–15.
18. Bolce, M. E., Hemmati-Brivanlou, A., Kushner, P. D., and Harland, R. M. (1992). *Development*, **115**, 681–8.
19. Kintner, C. R. and Brockes, J. P. (1984). *Nature*, **308**, 67–9.
20. Dale, L., Smith, J. C., and Slack, J. M. W. (1985). *J. Embryol. Exp. Morph.*, **89**, 289–312.
21. Cooke, J. and Smith, J. C. (1987). *Development*, **99**, 197–210.
22. Godsave, S. F., Anderton, B. H., and Wylie, C. C. (1986). *J. Embryol. Exp. Morph.*, **97**, 201–23.
23. Jones, E. A. and Woodland, H. R. (1989). *Development*, **107**, 785–91.

24. Hausen, P. and Dreyer, C. (1981). *Stain Technol.*, **56**, 287–93.
25. Harlow, E. and Lane, D. (1988). *Antibodies. A laboratory manual.* Cold Spring Harbor Press, New York.
26. Klymkowsky, M. W. and Hanken, J. (1991). *Meth. Cell Biol.*, **36**, 419–41.
27. Sambrook, J., Fritsch, E. F., and Maniatis, T. (1989). *Molecular cloning. A laboratory manual.* Cold Spring Harbor Press, New York.
28. Cooke, J., Smith, J. C., Smith, E. J., and Yaqoob, M. (1987). *Development,* **101,** 893–908.
29. Krieg, P. A. and Melton, D. A. (1984). *Nucl. Acids Res.*, **12**, 7057–70.
30. Colman, A. (1984). In *Transcription and translation: a practical approach* (ed. B. D. Hames and S. J. Higgins), pp. 271–302. IRL Press, Oxford.
31. Kay, B. K. (1991). *Meth. Cell Biol.*, **36**, 663–9.
32. McMahon, A. P. and Moon, R. T. (1989). *Cell,* **58,** 1075–84.
33. Dale, L., Howes, G., Price, B. M. J., and Smith, J. C. (1992). *Development,* **115,** 573–85.
34. Jones, C. M., Lyons, K. M., Lapan, P. M., Wright, C. V. E., and Hogan, B. L. M. (1992). *Development,* **115,** 639–47.
35. Cunliffe, V. and Smith, J. C. (1992). *Nature,* **358,** 427–30.
36. Kao, K. R. and Elinson, R. P. (1988). *Dev. Biol.,* **127,** 64–77.
37. Kao, K. and Danilchik, M. (1991). *Meth. Cell Biol.,* **36,** 271–84.
38. Smith, W. C. and Harland, R. M. (1992). *Cell,* **70,** 829–40.
39. Cho, K. W. Y., Blumberg, B., Steinbesser, H., and DeRobertis, E. M. (1991). *Cell,* **67,** 1111–20.
40. Smith, J. C., Price, B. M. J., Green, J. B. A., Weigel, D., and Hermann, B. G. (1991). *Cell,* **67,** 79–87.
41. Mohun, T. J., Brennan, S., Dathan, N., Fairman, S., and Gurdon, J. B. (1984). *Nature,* **311,** 716–21.
42. Christian, J. L., McMahon, J. A., McMahon, A. P., and Moon, R. T. (1991). *Development,* **111,** 1045–55.
43. Ruiz i Altaba, A. and Melton, D. A. (1989). *Development,* **106,** 173–83.
44. Wright, C. V. E., Morita, E. A., Wilkin, D. J., and DeRobertis, E. M. (1990). *Development,* **109,** 225–34.

A

Suppliers of specialist items

Atochem, La Défense 10, 428 Cours Michelet, F-92800 Puteaux, Hauts-de-Seine, France. *In UK*, Atochem UK Ltd., Colthrio Lane, Thatcham, Newbury, Berkshire RG13 4NR, UK. *In USA*, Atochem USA, 266 Harristown Road, PO Box 607, GlenRock, NJ 07452, USA.

Beckman Instruments Ltd, 45 Belmont Drive, PO Box 6764, Somerset, NK 08875-6764, USA. *In Europe*, Beckman Instruments SA., 22 Rue Juste-Olivier, CH-1260 Nyon, Switzerland.

Bio-Rad Laboratories, 1414 Harbor Way, Richmond, CA 94084, USA. *In UK*, Bio-Rad Laboratories Ltd, Bio-Rad House, Maylands Avenue, Hemel Hempstead, Herts HP2 7TD, UK.

Boehringer–Mannheim Biochemicals, Sandhoferstrasse 116, Postfach 310120, D6800 Mannheim, Germany. *In USA*, Boehringer–Mannheim, PO Box 50414, Indianapolis, IN 46250, USA. *In UK*, BCL-Boehringer, Boehringer–Mannheim House, Bell Lane, Lewes, East Sussex BN7 1LG, UK.

B. Braun Biotech, Inc., 999 Postal Road, Allentown, PA 18103, USA. *In Europe*, B. Braun Melsungen, Postfach 110, Carl-Braun Straße, 1-3508 Melsungen, Germany.

Brownlea Labs Inc., 2045 Martin Ave., Santa Clara, CA 95050, USA.

Campden Instruments, King Street, Sileby, Loughborough LE12 7LZ, UK.

Citifluor Ltd, Connaught Building, City University, Northampton Square, London EC1 0HB, UK.

Cole-Parmer Instrument Co., 7425 N. Osk Park Avenue, Chicago, IL 60648, USA.

Collaborative Research, PO Box 370068, Boston, MA 02241, USA.

Diagen GmbH, Max-Volmer Straße 4, 4010 Hilden, Germany. *In USA*, Qiagen Inc., 9259 Eton Avenue, Chatsworth, CA 91311, USA.

Dowex see Bio-Rad.

Du Pont Company/Biotechnology Systems Division, Barley Mill Plaza, Wilmington, DE 19805, USA. *In UK*, DuPont (UK) Ltd, Biotechnology Systems, Wedwood Way, Stevenage, Herts SG1 4QN, UK.

Fluka Chemie AG, Industriestrasse 25, CH-9470 Buchs, Switzerland. *In USA*, Fluka Chemical Corp., 980 South 2nd Street, Ronkonkoma, NY11779, USA. *In UK*, Fluka Chemicals, Ltd, Peakdale Road, Glossop, Derbyshire, UK.

GIBCO aka Life Technologies, PO Box 68, Grand Island, NY 14702, USA. *In UK,* GIBCO–BRL aka Life Technologies Ltd, Unit 4, Coeley Mill Trading Estate, Longbridge Way, Uxbridge UB8 2YG, UK.

Goodfellow Metals Ltd, Cambridge Science Park, Milton Road, Cambridge CB4 4DJ, UK.

Hilgenberg GmbH, Strauchgraben 2, PO Box 1161, D-3509 Malsfeld, Germany.

ICN Biomedicals, Inc., 3300 Hyland Avenue, Costa Mesa, CA 92626, USA. *In UK,* ICN Biomedicals Ltd, Lincoln Road, Cressey Industrial Estate, High Wycombe, Bucks HP12 3XJ, UK.

Inolex Chemical Company, Jackson and Swanson Streets, Philadelphia, PA 19148, USA.

LAUDA, Dr R. Wobser GmbH and Co. KG, Postfach 1251, D-6970 Lauda-Konigshofen, Germany.

Life Sciences, Inc., 1818 Market Street, Philadelphia PA 19103, USA.

Micro Instruments, 7 Little Clarendon Street, Oxford, UK.

Millipore UK Ltd, The Boulevard, Blackmoor Lane, Watford, Herts WD1 8YW, UK.

Monsanto Co., 800 North Lindbergh Boulevard, St Louis, MO 63167, USA.

New England Biolabs, Inc., 32 Tozer Road, Beverley, MA 01915, USA. *In Europe,* New England Biolabs GmbH, Postfach 2750, D-6231.

New England Nuclear, 549 Albany Street, Boston, MA 02118, USA. *In Europe*, du Pont de Nemours (Deutschland) GmbH, Biotechnology Systems Division, NEN Research Products, Postfach 40 12 40, 6072 Dreieich 4, German. (*See also* DuPont).

Nunc A/S, Postbox 280, Kampstrup, DK 4000, Roskilde, Denmark. *In UK,* Life Technologies Ltd, PO Box 35, Trident House, Renfrew Road, Paisley, PA3 4EF, UK. *In USA* Nunc Inc., 2000 North Aurora Road, Naperville, IL 60566, USA.

Pharmacia LKB Biotechnology AB, S-75182, Uppsala, Sweden. *In USA,* Pharmacia LKB Biotechnology Inc., 800 Centennial Avenue, Piscatawy, NJ 08855, USA.

Pierce, PO Box 117, Rockford, IL 61105, USA.

Promega Corp., 2800 Woods Hollow Road, Madison, WI 53711, USA.

Qiagen see Diagen.

Sigma Chemical Co., PO Box 14508, St. Louis, MO 63178, USA. *In UK* Sigma Chemical Co. Ltd, Fancy Road, Poole, Dorset BH17 7NH, UK.

Singer Instruments Co., Roadwater, Watchet, Somerset TA23 0QL, UK.

Sorall see Du Pont.

UV Light Products, Inc., 5100 Walnut Grove Avenue, PO Box 1501, San Gabriel, CA 91778, USA. *In UK,* UVP Ltd, Science Park, Milton Road, Cambridge CB4 4BN, UK.

Whatman Biosystems Inc., 22 Bridewell Place, Clifton, NJ 07014, USA. *In UK*, Whatman Biosystems Ltd, Spigfield Mill, Maidstone, Kent ME14 2LE, UK.

Index

ORDER OTHER TITLES OF INTEREST TODAY

Price list for: UK, Europe, Rest of World (excluding US and Canada)

Forthcoming Titles

124. **Human Genetic Disease Analysis** Davies, K.E. (Ed)
...... Spiralbound hardback 0-19-963309-6 **£30.00**
...... Paperback 0-19-963308-8 **£18.50**
123. **Protein Phosphorylation** Hardie, G. (Ed)
...... Spiralbound hardback 0-19-963306-1 **£32.50**
...... Paperback 0-19-963305-3 **£22.50**
122. **Immunocytochemistry** Beesley, J. (Ed)
...... Spiralbound hardback 0-19-963270-7 **£32.50**
...... Paperback 0-19-963269-3 **£22.50**
121. **Tumour Immunobiology** Gallagher, G., Rees, R.C. & others (Eds)
...... Spiralbound hardback 0-19-963370-3 **£35.00**
...... Paperback 0-19-963369-X **£25.00**
120. **Transcription Factors** Latchman, D.S. (Ed)
...... Spiralbound hardback 0-19-963342-8 **£30.00**
...... Paperback 0-19-963341-X **£19.50**
119. **Growth Factors** McKay, I.A. & Leigh, I. (Eds)
...... Spiralbound hardback 0-19-963360-6 **£30.00**
...... Paperback 0-19-963359-2 **£19.50**
118. **Histocompatibility Testing** Dyer, P. & Middleton, D. (Eds)
...... Spiralbound hardback 0-19-963364-9 **£32.50**
...... Paperback 0-19-963363-0 **£22.50**
117. **Gene Transcription** Hames, D.B. & Higgins, S.J. (Eds)
...... Spiralbound hardback 0-19-963292-8 **£35.00**
...... Paperback 0-19-963291-X **£25.00**
116. **Electrophysiology** Wallis, D.I. (Ed)
...... Spiralbound hardback 0-19-963348-7 **£32.50**
...... Paperback 0-19-963347-9 **£22.50**
115. **Biological Data Analysis** Fry, J.C. (Ed)
...... Spiralbound hardback 0-19-963340-1 **£50.00**
...... Paperback 0-19-963339-8 **£27.50**
114. **Experimental Neuroanatomy** Bolam, J.P. (Ed)
...... Spiralbound hardback 0-19-963326-6 **£32.50**
...... Paperback 0-19-963325-8 **£22.50**
112. **Lipid Analysis** Hamilton, R.J. & Hamilton, S.J. (Eds)
...... Spiralbound hardback 0-19-963098-4 **£35.00**
...... Paperback 0-19-963099-2 **£25.00**
111. **Haemopoiesis** Testa, N.G. & Molineux, G. (Eds)
...... Spiralbound hardback 0-19-963366-5 **£32.50**
...... Paperback 0-19-963365-7 **£22.50**

Published Titles

113. **Preparative Centrifugation** Rickwood, D. (Ed)
...... Spiralbound hardback 0-19-963208-1 **£45.00**
...... Paperback 0-19-963211-1 **£25.00**
110. **Pollination Ecology** Dafni, A.
...... Spiralbound hardback 0-19-963299-5 **£32.50**
...... Paperback 0-19-963298-7 **£22.50**
109. **In Situ Hybridization** Wilkinson, D.G. (Ed)
...... Spiralbound hardback 0-19-963328-2 **£30.00**
...... Paperback 0-19-963327-4 **£18.50**
108. **Protein Engineering** Rees, A.R., Sternberg, M.J.E. & others (Eds)
...... Spiralbound hardback 0-19-963139-5 **£35.00**
...... Paperback 0-19-963138-7 **£25.00**

107. **Cell-Cell Interactions** Stevenson, B.R., Gallin, W.J. & others (Eds)
...... Spiralbound hardback 0-19-963319-3 **£32.50**
...... Paperback 0-19-963318-5 **£22.50**
106. **Diagnostic Molecular Pathology: Volume I** Herrington, C.S. & McGee, J. O'D. (Eds)
...... Spiralbound hardback 0-19-963237-5 **£30.00**
...... Paperback 0-19-963236-7 **£19.50**
105. **Biomechanics-Materials** Vincent, J.F.V. (Ed)
...... Spiralbound hardback 0-19-963223-5 **£35.00**
...... Paperback 0-19-963222-7 **£25.00**
104. **Animal Cell Culture (2/e)** Freshney, R.I. (Ed)
...... Spiralbound hardback 0-19-963212-X **£30.00**
...... Paperback 0-19-963213-8 **£19.50**
103. **Molecular Plant Pathology: Volume II** Gurr, S.J., McPherson, M.J. & others (Eds)
...... Spiralbound hardback 0-19-963352-5 **£32.50**
...... Paperback 0-19-963351-7 **£22.50**
101. **Protein Targeting** Magee, A.I. & Wileman, T. (Eds)
...... Spiralbound hardback 0-19-963206-5 **£32.50**
...... Paperback 0-19-963210-3 **£22.50**
100. **Diagnostic Molecular Pathology: Volume II: Cell and Tissue Genotyping** Herrington, C.S. & McGee, J.O'D. (Eds)
...... Spiralbound hardback 0-19-963239-1 **£30.00**
...... Paperback 0-19-963238-3 **£19.50**
99. **Neuronal Cell Lines** Wood, J.N. (Ed)
...... Spiralbound hardback 0-19-963346-0 **£32.50**
...... Paperback 0-19-963345-2 **£22.50**
98. **Neural Transplantation** Dunnett, S.B. & Björklund, A. (Eds)
...... Spiralbound hardback 0-19-963286-3 **£30.00**
...... Paperback 0-19-963285-5 **£19.50**
97. **Human Cytogenetics: Volume II: Malignancy and Acquired Abnormalities (2/e)** Rooney, D.E. & Czepulkowski, B.H. (Eds)
...... Spiralbound hardback 0-19-963290-1 **£30.00**
...... Paperback 0-19-963289-8 **£22.50**
96. **Human Cytogenetics: Volume I: Constitutional Analysis (2/e)** Rooney, D.E. & Czepulkowski, B.H. (Eds)
...... Spiralbound hardback 0-19-963288-X **£30.00**
...... Paperback 0-19-963287-1 **£22.50**
95. **Lipid Modification of Proteins** Hooper, N.M. & Turner, A.J. (Eds)
...... Spiralbound hardback 0-19-963274-X **£32.50**
...... Paperback 0-19-963273-1 **£22.50**
94. **Biomechanics-Structures and Systems** Biewener, A.A. (Ed)
...... Spiralbound hardback 0-19-963268-5 **£42.50**
...... Paperback 0-19-963267-7 **£25.00**
93. **Lipoprotein Analysis** Converse, C.A. & Skinner, E.R. (Eds)
...... Spiralbound hardback 0-19-963192-1 **£30.00**
...... Paperback 0-19-963231-6 **£19.50**
92. **Receptor-Ligand Interactions** Hulme, E.C. (Ed)
...... Spiralbound hardback 0-19-963090-9 **£35.00**
...... Paperback 0-19-963091-7 **£25.00**
91. **Molecular Genetic Analysis of Populations** Hoelzel, A.R. (Ed)
...... Spiralbound hardback 0-19-963278-2 **£32.50**
...... Paperback 0-19-963277-4 **£22.50**

90. **Enzyme Assays** Eisenthal, R. & Danson, M.J. (Eds)
.... Spiralbound hardback 0-19-963142-5 **£35.00**
.... Paperback 0-19-963143-3 **£25.00**
89. **Microcomputers in Biochemistry** Bryce, C.F.A. (Ed)
.... Spiralbound hardback 0-19-963253-7 **£30.00**
.... Paperback 0-19-963252-9 **£19.50**
88. **The Cytoskeleton** Carraway, K.L. & Carraway, C.A.C. (Eds)
.... Spiralbound hardback 0-19-963257-X **£30.00**
.... Paperback 0-19-963256-1 **£19.50**
87. **Monitoring Neuronal Activity** Stamford, J.A. (Ed)
.... Spiralbound hardback 0-19-963244-8 **£30.00**
.... Paperback 0-19-963243-X **£19.50**
86. **Crystallization of Nucleic Acids and Proteins** Ducruix, A. & Gieg‹130›, R. (Eds)
.... Spiralbound hardback 0-19-963245-6 **£35.00**
.... Paperback 0-19-963246-4 **£25.00**
85. **Molecular Plant Pathology: Volume I** Gurr, S.J., McPherson, M.J. & others (Eds)
.... Spiralbound hardback 0-19-963103-4 **£30.00**
.... Paperback 0-19-963102-6 **£19.50**
84. **Anaerobic Microbiology** Levett, P.N. (Ed)
.... Spiralbound hardback 0-19-963204-9 **£32.50**
.... Paperback 0-19-963262-6 **£22.50**
83. **Oligonucleotides and Analogues** Eckstein, F. (Ed)
.... Spiralbound hardback 0-19-963280-4 **£32.50**
.... Paperback 0-19-963279-0 **£22.50**
82. **Electron Microscopy in Biology** Harris, R. (Ed)
.... Spiralbound hardback 0-19-963219-7 **£32.50**
.... Paperback 0-19-963215-4 **£22.50**
81. **Essential Molecular Biology: Volume II** Brown, T.A. (Ed)
.... Spiralbound hardback 0-19-963112-3 **£32.50**
.... Paperback 0-19-963113-1 **£22.50**
80. **Cellular Calcium** McCormack, J.G. & Cobbold, P.H. (Eds)
.... Spiralbound hardback 0-19-963131-X **£35.00**
.... Paperback 0-19-963130-1 **£25.00**
79. **Protein Architecture** Lesk, A.M.
.... Spiralbound hardback 0-19-963054-2 **£32.50**
.... Paperback 0-19-963055-0 **£22.50**
78. **Cellular Neurobiology** Chad, J. & Wheal, H. (Eds)
.... Spiralbound hardback 0-19-963106-9 **£32.50**
.... Paperback 0-19-963107-7 **£22.50**
77. **PCR** McPherson, M.J., Quirke, P. & others (Eds)
.... Spiralbound hardback 0-19-963226-X **£30.00**
.... Paperback 0-19-963196-4 **£19.50**
76. **Mammalian Cell Biotechnology** Butler, M. (Ed)
.... Spiralbound hardback 0-19-963207-3 **£30.00**
.... Paperback 0-19-963209-X **£19.50**
75. **Cytokines** Balkwill, F.R. (Ed)
.... Spiralbound hardback 0-19-963218-9 **£35.00**
.... Paperback 0-19-963214-6 **£25.00**
74. **Molecular Neurobiology** Chad, J. & Wheal, H. (Eds)
.... Spiralbound hardback 0-19-963108-5 **£30.00**
.... Paperback 0-19-963109-3 **£19.50**
73. **Directed Mutagenesis** McPherson, M.J. (Ed)
.... Spiralbound hardback 0-19-963141-7 **£30.00**
.... Paperback 0-19-963140-9 **£19.50**
72. **Essential Molecular Biology: Volume I** Brown, T.A. (Ed)
.... Spiralbound hardback 0-19-963110-7 **£32.50**
.... Paperback 0-19-963111-5 **£22.50**
71. **Peptide Hormone Action** Siddle, K. & Hutton, J.C.
.... Spiralbound hardback 0-19-963070-4 **£32.50**
.... Paperback 0-19-963071-2 **£22.50**
70. **Peptide Hormone Secretion** Hutton, J.C. & Siddle, K. (Eds)
.... Spiralbound hardback 0-19-963068-2 **£35.00**
.... Paperback 0-19-963069-0 **£25.00**
69. **Postimplantation Mammalian Embryos** Copp, A.J. & Cockroft, D.L. (Eds)
.... Spiralbound hardback 0-19-963088-7 **£35.00**
.... Paperback 0-19-963089-5 **£25.00**
68. **Receptor-Effector Coupling** Hulme, E.C. (Ed)
.... Spiralbound hardback 0-19-963094-1 **£30.00**
.... Paperback 0-19-963095-X **£19.50**

67. **Gel Electrophoresis of Proteins (2/e)** Hames, B.D. & Rickwood, D. (Eds)
.... Spiralbound hardback 0-19-963074-7 **£35.00**
.... Paperback 0-19-963075-5 **£25.00**
66. **Clinical Immunology** Gooi, H.C. & Chapel, H. (Eds)
.... Spiralbound hardback 0-19-963086-0 **£32.50**
.... Paperback 0-19-963087-9 **£22.50**
65. **Receptor Biochemistry** Hulme, E.C. (Ed)
.... Spiralbound hardback 0-19-963092-5 **£35.00**
.... Paperback 0-19-963093-3 **£25.00**
64. **Gel Electrophoresis of Nucleic Acids (2/e)** Rickwood, D. & Hames, B.D. (Eds)
.... Spiralbound hardback 0-19-963082-8 **£32.50**
.... Paperback 0-19-963083-6 **£22.50**
63. **Animal Virus Pathogenesis** Oldstone, M.B.A. (Ed)
.... Spiralbound hardback 0-19-963100-X **£30.00**
.... Paperback 0-19-963101-8 **£18.50**
62. **Flow Cytometry** Ormerod, M.G. (Ed)
.... Paperback 0-19-963053-4 **£22.50**
61. **Radioisotopes in Biology** Slater, R.J. (Ed)
.... Spiralbound hardback 0-19-963080-1 **£32.50**
.... Paperback 0-19-963081-X **£22.50**
60. **Biosensors** Cass, A.E.G. (Ed)
.... Spiralbound hardback 0-19-963046-1 **£30.00**
.... Paperback 0-19-963047-X **£19.50**
59. **Ribosomes and Protein Synthesis** Spedding, G. (Ed)
.... Spiralbound hardback 0-19-963104-2 **£32.50**
.... Paperback 0-19-963105-0 **£22.50**
58. **Liposomes** New, R.R.C. (Ed)
.... Spiralbound hardback 0-19-963076-3 **£35.00**
.... Paperback 0-19-963077-1 **£22.50**
57. **Fermentation** McNeil, B. & Harvey, L.M. (Eds)
.... Spiralbound hardback 0-19-963044-5 **£30.00**
.... Paperback 0-19-963045-3 **£19.50**
56. **Protein Purification Applications** Harris, E.L.V. & Angal, S. (Eds)
.... Spiralbound hardback 0-19-963022-4 **£30.00**
.... Paperback 0-19-963023-2 **£18.50**
55. **Nucleic Acids Sequencing** Howe, C.J. & Ward, E.S. (Eds)
.... Spiralbound hardback 0-19-963056-9 **£30.00**
.... Paperback 0-19-963057-7 **£19.50**
54. **Protein Purification Methods** Harris, E.L.V. & Angal, S. (Eds)
.... Spiralbound hardback 0-19-963002-X **£30.00**
.... Paperback 0-19-963003-8 **£20.00**
53. **Solid Phase Peptide Synthesis** Atherton, E. & Sheppard, R.C.
.... Spiralbound hardback 0-19-963066-6 **£30.00**
.... Paperback 0-19-963067-4 **£18.50**
52. **Medical Bacteriology** Hawkey, P.M. & Lewis, D.A. (Eds)
.... Spiralbound hardback 0-19-963008-9 **£38.00**
.... Paperback 0-19-963009-7 **£25.00**
51. **Proteolytic Enzymes** Beynon, R.J. & Bond, J.S. (Eds)
.... Spiralbound hardback 0-19-963058-5 **£30.00**
.... Paperback 0-19-963059-3 **£19.50**
50. **Medical Mycology** Evans, E.G.V. & Richardson, M.D. (Eds)
.... Spiralbound hardback 0-19-963010-0 **£37.50**
.... Paperback 0-19-963011-9 **£25.00**
49. **Computers in Microbiology** Bryant, T.N. & Wimpenny, J.W.T. (Eds)
.... Paperback 0-19-963015-1 **£19.50**
48. **Protein Sequencing** Findlay, J.B.C. & Geisow, M.J. (Eds)
.... Spiralbound hardback 0-19-963012-7 **£30.00**
.... Paperback 0-19-963013-5 **£18.50**
47. **Cell Growth and Division** Baserga, R. (Ed)
.... Spiralbound hardback 0-19-963026-7 **£30.00**
.... Paperback 0-19-963027-5 **£18.50**
46. **Protein Function** Creighton, T.E. (Ed)
.... Spiralbound hardback 0-19-963006-2 **£32.50**
.... Paperback 0-19-963007-0 **£22.50**
45. **Protein Structure** Creighton, T.E. (Ed)
.... Spiralbound hardback 0-19-963000-3 **£32.50**
.... Paperback 0-19-963001-1 **£22.50**
44. **Antibodies: Volume II** Catty, D. (Ed)
.... Spiralbound hardback 0-19-963018-6 **£30.00**
.... Paperback 0-19-963019-4 **£19.50**

43.	**HPLC of Macromolecules** Oliver, R.W.A. (Ed)		
......	Spiralbound hardback	0-19-963020-8	**£30.00**
......	Paperback	0-19-963021-6	**£19.50**
42.	**Light Microscopy in Biology** Lacey, A.J. (Ed)		
......	Spiralbound hardback	0-19-963036-4	**£30.00**
......	Paperback	0-19-963037-2	**£19.50**
41.	**Plant Molecular Biology** Shaw, C.H. (Ed)		
......	Paperback	1-85221-056-7	**£22.50**
40.	**Microcomputers in Physiology** Fraser, P.J. (Ed)		
......	Spiralbound hardback	1-85221-129-6	**£30.00**
......	Paperback	1-85221-130-X	**£19.50**
39.	**Genome Analysis** Davies, K.E. (Ed)		
......	Spiralbound hardback	1-85221-109-1	**£30.00**
......	Paperback	1-85221-110-5	**£18.50**
38.	**Antibodies: Volume I** Catty, D. (Ed)		
......	Paperback	0-947946-85-3	**£19.50**
37.	**Yeast** Campbell, I. & Duffus, J.H. (Eds)		
......	Paperback	0-947946-79-9	**£19.50**
36.	**Mammalian Development** Monk, M. (Ed)		
......	Hardback	1-85221-030-3	**£30.50**
......	Paperback	1-85221-029-X	**£22.50**
35.	**Lymphocytes** Klaus, G.G.B. (Ed)		
......	Hardback	1-85221-018-4	**£30.00**
34.	**Lymphokines and Interferons** Clemens, M.J., Morris, A.G. & others (Eds)		
......	Paperback	1-85221-035-4	**£22.50**
33.	**Mitochondria** Darley-Usmar, V.M., Rickwood, D. & others (Eds)		
......	Hardback	1-85221-034-6	**£32.50**
......	Paperback	1-85221-033-8	**£22.50**
32.	**Prostaglandins and Related Substances** Benedetto, C., McDonald-Gibson, R.G. & others (Eds)		
......	Hardback	1-85221-032-X	**£32.50**
......	Paperback	1-85221-031-1	**£22.50**
31.	**DNA Cloning: Volume III** Glover, D.M. (Ed)		
......	Hardback	1-85221-049-4	**£30.00**
......	Paperback	1-85221-048-6	**£19.50**
30.	**Steroid Hormones** Green, B. & Leake, R.E. (Eds)		
......	Paperback	0-947946-53-5	**£19.50**
29.	**Neurochemistry** Turner, A.J. & Bachelard, H.S. (Eds)		
......	Hardback	1-85221-028-1	**£30.00**
......	Paperback	1-85221-027-3	**£19.50**
28.	**Biological Membranes** Findlay, J.B.C. & Evans, W.H. (Eds)		
......	Hardback	0-947946-84-5	**£32.50**
......	Paperback	0-947946-83-7	**£22.50**
27.	**Nucleic Acid and Protein Sequence Analysis** Bishop, M.J. & Rawlings, C.J. (Eds)		
......	Hardback	1-85221-007-9	**£35.00**
......	Paperback	1-85221-006-0	**£25.00**
26.	**Electron Microscopy in Molecular Biology** Sommerville, J. & Scheer, U. (Eds)		
......	Hardback	0-947946-64-0	**£30.00**
......	Paperback	0-947946-54-3	**£19.50**
25.	**Teratocarcinomas and Embryonic Stem Cells** Robertson, E.J. (Ed)		
......	Hardback	1-85221-005-2	**£19.50**
......	Paperback	1-85221-004-4	**£19.50**
24.	**Spectrophotometry and Spectrofluorimetry** Harris, D.A. & Bashford, C.L. (Eds)		
......	Hardback	0-947946-69-1	**£30.00**
......	Paperback	0-947946-46-2	**£18.50**
23.	**Plasmids** Hardy, K.G. (Ed)		
......	Paperback	0-947946-81-0	**£18.50**
22.	**Biochemical Toxicology** Snell, K. & Mullock, B. (Eds)		
......	Paperback	0-947946-52-7	**£19.50**
19.	**Drosophila** Roberts, D.B. (Ed)		
......	Hardback	0-947946-66-7	**£32.50**
......	Paperback	0-947946-45-4	**£22.50**
17.	**Photosynthesis: Energy Transduction** Hipkins, M.F. & Baker, N.R. (Eds)		
......	Hardback	0-947946-63-2	**£30.00**
......	Paperback	0-947946-51-9	**£18.50**
16.	**Human Genetic Diseases** Davies, K.E. (Ed)		
......	Hardback	0-947946-76-4	**£30.00**
......	Paperback	0-947946-75-6	**£18.50**

14.	**Nucleic Acid Hybridisation** Hames, B.D. & Higgins, S.J. (Eds)		
......	Hardback	0-947946-61-6	**£30.00**
......	Paperback	0-947946-23-3	**£19.50**
13.	**Immobilised Cells and Enzymes** Woodward, J. (Ed)		
......	Hardback	0-947946-60-8	**£18.50**
12.	**Plant Cell Culture** Dixon, R.A. (Ed)		
......	Paperback	0-947946-22-5	**£19.50**
11a.	**DNA Cloning: Volume I** Glover, D.M. (Ed)		
......	Paperback	0-947946-18-7	**£18.50**
11b.	**DNA Cloning: Volume II** Glover, D.M. (Ed)		
......	Paperback	0-947946-19-5	**£19.50**
10.	**Virology** Mahy, B.W.J. (Ed)		
......	Paperback	0-904147-78-9	**£19.50**
9.	**Affinity Chromatography** Dean, P.D.G., Johnson, W.S. & others (Eds)		
......	Paperback	0-904147-71-1	**£19.50**
7.	**Microcomputers in Biology** Ireland, C.R. & Long, S.P. (Eds)		
......	Paperback	0-904147-57-6	**£18.00**
6.	**Oligonucleotide Synthesis** Gait, M.J. (Ed)		
......	Paperback	0-904147-74-6	**£18.50**
5.	**Transcription and Translation** Hames, B.D. & Higgins, S.J. (Eds)		
......	Paperback	0-904147-52-5	**£22.50**
3.	**Iodinated Density Gradient Media** Rickwood, D. (Ed)		
......	Paperback	0-904147-51-7	**£19.50**

Sets

	Essential Molecular Biology: Volumes I and II as a set Brown, T.A. (Ed)		
......	Spiralbound hardback	0-19-963114-X	**£58.00**
......	Paperback	0-19-963115-8	**£40.00**
	Antibodies: Volumes I and II as a set Catty, D. (Ed)		
......	Paperback	0-19-963063-1	**£33.00**
	Cellular and Molecular Neurobiology Chad, J. & Wheal, H. (Eds)		
......	Spiralbound hardback	0-19-963255-3	**£56.00**
......	Paperback	0-19-963254-5	**£38.00**
	Protein Structure and Protein Function: Two-volume set Creighton, T.E. (Ed)		
......	Spiralbound hardback	0-19-963064-X	**£55.00**
......	Paperback	0-19-963065-8	**£38.00**
	DNA Cloning: Volumes I, II, III as a set Glover, D.M. (Ed)		
......	Paperback	1-85221-069-9	**£46.00**
	Molecular Plant Pathology: Volumes I and II as a set Gurr, S.J., McPherson, M.J. & others (Eds)		
......	Spiralbound hardback	0-19-963354-1	**£56.00**
......	Paperback	0-19-963353-3	**£37.00**
	Protein Purification Methods, and Protein Purification Applications, two-volume set Harris, E.L.V. & Angal, S. (Eds)		
......	Spiralbound hardback	0-19-963048-8	**£48.00**
......	Paperback	0-19-963049-6	**£32.00**
	Diagnostic Molecular Pathology: Volumes I and II as a set Herrington, C.S. & McGee, J. O'D. (Eds)		
......	Spiralbound hardback	0-19-963241-3	**£54.00**
......	Paperback	0-19-963240-5	**£35.00**
	Receptor Biochemistry; Receptor-Effector Coupling; Receptor-Ligand Interactions Hulme, E.C. (Ed)		
......	Spiralbound hardback	0-19-963096-8	**£90.00**
......	Paperback	0-19-963097-6	**£62.50**
	Signal Transduction Milligan, G. (Ed)		
......	Spiralbound hardback	0-19-963296-0	**£30.00**
......	Paperback	0-19-963295-2	**£18.50**
	Human Cytogenetics: Volumes I and II as a set (2/e) Rooney, D.E. & Czepulkowski, B.H. (Eds)		
......	Hardback	0-19-963314-2	**£58.50**
......	Paperback	0-19-963313-4	**£40.50**
	Peptide Hormone Secretion/Peptide Hormone Action Siddle, K. & Hutton, J.C. (Eds)		
......	Spiralbound hardback	0-19-963072-0	**£55.00**
......	Paperback	0-19-963073-9	**£38.00**

ORDER FORM for UK, Europe and Rest of World

(Excluding USA and Canada)

Qty	ISBN	Author	Title	Amount
			P&P	
			TOTAL	

Please add postage and packing: £1.75 for UK orders under £20; £2.75 for UK orders over £20; overseas orders add 10% of total.

Name ....

Address ....

....

.... Post code ....

[ ] Please charge £ .... to my credit card
Access/VISA/Eurocard/AMEX/Diners Club (circle appropriate card)

Card No .... Expiry date ....

Signature ....

Credit card account address if different from above:

....

.... Postcode ....

[ ] I enclose a cheque for £.....................

Please return this form to: OUP Distribution Services, Saxon Way West, Corby, Northants NN18 9ES

OR ORDER BY CREDIT CARD HOTLINE: Tel +44-(0)536-741519 or Fax +44-(0)536-746337

ORDER OTHER TITLES OF INTEREST TODAY

Price list for: USA and Canada

123. **Protein Phosphorylation** Hardie, G. (Ed)
...... Spiralbound hardback 0-19-963306-1 **$65.00**
...... Paperback 0-19-963305-3 **$45.00**
121. **Tumour Immunobiology** Gallagher, G., Rees, R.C. & others (Eds)
...... Spiralbound hardback 0-19-963370-3 **$72.00**
...... Paperback 0-19-963369-X **$50.00**
117. **Gene Transcription** Hames, D.B. & Higgins, S.J. (Eds)
...... Spiralbound hardback 0-19-963292-8 **$72.00**
...... Paperback 0-19-963291-X **$50.00**
116. **Electrophysiology** Wallis, D.I. (Ed)
...... Spiralbound hardback 0-19-963348-7 **$66.50**
...... Paperback 0-19-963347-9 **$45.95**
115. **Biological Data Analysis** Fry, J.C. (Ed)
...... Spiralbound hardback 0-19-963340-1 **$80.00**
...... Paperback 0-19-963339-8 **$60.00**
114. **Experimental Neuroanatomy** Bolam, J.P. (Ed)
...... Spiralbound hardback 0-19-963326-6 **$65.00**
...... Paperback 0-19-963325-8 **$40.00**
111. **Haemopoiesis** Testa, N.G. & Molineux, G. (Eds)
...... Spiralbound hardback 0-19-963366-5 **$65.00**
...... Paperback 0-19-963365-7 **$45.00**
113. **Preparative Centrifugation** Rickwood, D. (Ed)
...... Spiralbound hardback 0-19-963208-1 **$90.00**
...... Paperback 0-19-963211-1 **$50.00**
110. **Pollination Ecology** Dafni, A.
...... Spiralbound hardback 0-19-963299-5 **$65.00**
...... Paperback 0-19-963298-7 **$45.00**
109. **In Situ Hybridization** Wilkinson, D.G. (Ed)
...... Spiralbound hardback 0-19-963328-2 **$58.00**
...... Paperback 0-19-963327-4 **$36.00**
108. **Protein Engineering** Rees, A.R., Sternberg, M.J.E. & others (Eds)
...... Spiralbound hardback 0-19-963139-5 **$75.00**
...... Paperback 0-19-963138-7 **$50.00**
107. **Cell-Cell Interactions** Stevenson, B.R., Gallin, W.J. & others (Eds)
...... Spiralbound hardback 0-19-963319-3 **$60.00**
...... Paperback 0-19-963318-5 **$40.00**
106. **Diagnostic Molecular Pathology: Volume I** Herrington, C.S. & McGee, J. O'D. (Eds)
...... Spiralbound hardback 0-19-963237-5 **$58.00**
...... Paperback 0-19-963236-7 **$38.00**
105. **Biomechanics-Materials** Vincent, J.F.V. (Ed)
...... Spiralbound hardback 0-19-963223-5 **$70.00**
...... Paperback 0-19-963222-7 **$50.00**
104. **Animal Cell Culture (2/e)** Freshney, R.I. (Ed)
...... Spiralbound hardback 0-19-963212-X **$60.00**
...... Paperback 0-19-963213-8 **$40.00**
103. **Molecular Plant Pathology: Volume II** Gurr, S.J., McPherson, M.J. & others (Eds)
...... Spiralbound hardback 0-19-963352-5 **$65.00**
...... Paperback 0-19-963351-7 **$45.00**
101. **Protein Targeting** Magee, A.I. & Wileman, T. (Eds)
...... Spiralbound hardback 0-19-963206-5 **$75.00**
...... Paperback 0-19-963210-3 **$50.00**
100. **Diagnostic Molecular Pathology: Volume II: Cell and Tissue Genotyping** Herrington, C.S. & McGee, J.O'D. (Eds)
...... Spiralbound hardback 0-19-963239-1 **$60.00**
...... Paperback 0-19-963238-3 **$39.00**

99. **Neuronal Cell Lines** Wood, J.N. (Ed)
...... Spiralbound hardback 0-19-963346-0 **$68.00**
...... Paperback 0-19-963345-2 **$48.00**
98. **Neural Transplantation** Dunnett, S.B. & Björklund, A. (Eds)
...... Spiralbound hardback 0-19-963286-3 **$69.00**
...... Paperback 0-19-963285-5 **$42.00**
97. **Human Cytogenetics: Volume II: Malignancy and Acquired Abnormalities (2/e)** Rooney, D.E. & Czepulkowski, B.H. (Eds)
...... Spiralbound hardback 0-19-963290-1 **$75.00**
...... Paperback 0-19-963289-8 **$50.00**
96. **Human Cytogenetics: Volume I: Constitutional Analysis (2/e)** Rooney, D.E. & Czepulkowski, B.H. (Eds)
...... Spiralbound hardback 0-19-963288-X **$75.00**
...... Paperback 0-19-963287-1 **$50.00**
95. **Lipid Modification of Proteins** Hooper, N.M. & Turner, A.J. (Eds)
...... Spiralbound hardback 0-19-963274-X **$75.00**
...... Paperback 0-19-963273-1 **$50.00**
94. **Biomechanics-Structures and Systems** Biewener, A.A. (Ed)
...... Spiralbound hardback 0-19-963268-5 **$85.00**
...... Paperback 0-19-963267-7 **$50.00**
93. **Lipoprotein Analysis** Converse, C.A. & Skinner, E.R. (Eds)
...... Spiralbound hardback 0-19-963192-1 **$65.00**
...... Paperback 0-19-963231-6 **$42.00**
92. **Receptor-Ligand Interactions** Hulme, E.C. (Ed)
...... Spiralbound hardback 0-19-963090-9 **$75.00**
...... Paperback 0-19-963091-7 **$50.00**
91. **Molecular Genetic Analysis of Populations** Hoelzel, A.R. (Ed)
...... Spiralbound hardback 0-19-963278-2 **$65.00**
...... Paperback 0-19-963277-4 **$45.00**
90. **Enzyme Assays** Eisenthal, R. & Danson, M.J. (Eds)
...... Spiralbound hardback 0-19-963142-5 **$68.00**
...... Paperback 0-19-963143-3 **$48.00**
89. **Microcomputers in Biochemistry** Bryce, C.F.A. (Ed)
...... Spiralbound hardback 0-19-963253-7 **$60.00**
...... Paperback 0-19-963252-9 **$40.00**
88. **The Cytoskeleton** Carraway, K.L. & Carraway, C.A.C. (Eds)
...... Spiralbound hardback 0-19-963257-X **$60.00**
...... Paperback 0-19-963256-1 **$40.00**
87. **Monitoring Neuronal Activity** Stamford, J.A. (Ed)
...... Spiralbound hardback 0-19-963244-8 **$60.00**
...... Paperback 0-19-963243-X **$40.00**
86. **Crystallization of Nucleic Acids and Proteins** Ducruix, A. & Gieg‹130›, R. (Eds)
...... Spiralbound hardback 0-19-963245-6 **$60.00**
...... Paperback 0-19-963246-4 **$50.00**
85. **Molecular Plant Pathology: Volume I** Gurr, S.J., McPherson, M.J. & others (Eds)
...... Spiralbound hardback 0-19-963103-4 **$60.00**
...... Paperback 0-19-963102-6 **$40.00**
84. **Anaerobic Microbiology** Levett, P.N. (Ed)
...... Spiralbound hardback 0-19-963204-9 **$75.00**
...... Paperback 0-19-963262-6 **$45.00**

83. **Oligonucleotides and Analogues** Eckstein, F.
(Ed)
...... Spiralbound hardback 0-19-963280-4 **$65.00**
...... Paperback 0-19-963279-0 **$45.00**
82. **Electron Microscopy in Biology** Harris, R. (Ed)
...... Spiralbound hardback 0-19-963219-7 **$65.00**
...... Paperback 0-19-963215-4 **$45.00**
81. **Essential Molecular Biology: Volume II** Brown,
T.A. (Ed)
...... Spiralbound hardback 0-19-963112-3 **$65.00**
...... Paperback 0-19-963113-1 **$45.00**
80. **Cellular Calcium** McCormack, J.G. & Cobbold,
P.H. (Eds)
...... Spiralbound hardback 0-19-963131-X **$75.00**
...... Paperback 0-19-963130-1 **$50.00**
79. **Protein Architecture** Lesk, A.M.
...... Spiralbound hardback 0-19-963054-2 **$65.00**
...... Paperback 0-19-963055-0 **$45.00**
78. **Cellular Neurobiology** Chad, J. & Wheal, H.
(Eds)
...... Spiralbound hardback 0-19-963106-9 **$73.00**
...... Paperback 0-19-963107-7 **$43.00**
77. **PCR** McPherson, M.J., Quirke, P. & others
(Eds)
...... Spiralbound hardback 0-19-963226-X **$55.00**
...... Paperback 0-19-963196-4 **$40.00**
76. **Mammalian Cell Biotechnology** Butler, M.
(Ed)
...... Spiralbound hardback 0-19-963207-3 **$60.00**
...... Paperback 0-19-963209-X **$40.00**
75. **Cytokines** Balkwill, F.R. (Ed)
...... Spiralbound hardback 0-19-963218-9 **$64.00**
...... Paperback 0-19-963214-6 **$44.00**
74. **Molecular Neurobiology** Chad, J. & Wheal, H.
(Eds)
...... Spiralbound hardback 0-19-963108-5 **$56.00**
...... Paperback 0-19-963109-3 **$36.00**
73. **Directed Mutagenesis** McPherson, M.J. (Ed)
...... Spiralbound hardback 0-19-963141-7 **$55.00**
...... Paperback 0-19-963140-9 **$35.00**
72. **Essential Molecular Biology: Volume I** Brown,
T.A. (Ed)
...... Spiralbound hardback 0-19-963110-7 **$65.00**
...... Paperback 0-19-963111-5 **$45.00**
71. **Peptide Hormone Action** Siddle, K. & Hutton,
J.C.
...... Spiralbound hardback 0-19-963070-4 **$70.00**
...... Paperback 0-19-963071-2 **$50.00**
70. **Peptide Hormone Secretion** Hutton, J.C. &
Siddle, K. (Eds)
...... Spiralbound hardback 0-19-963068-2 **$70.00**
...... Paperback 0-19-963069-0 **$50.00**
69. **Postimplantation Mammalian Embryos** Copp,
A.J. & Cockroft, D.L. (Eds)
...... Spiralbound hardback 0-19-963088-7 **$70.00**
...... Paperback 0-19-963089-5 **$50.00**
68. **Receptor-Effector Coupling** Hulme, E.C. (Ed)
...... Spiralbound hardback 0-19-963094-1 **$70.00**
...... Paperback 0-19-963095-X **$45.00**
67. **Gel Electrophoresis of Proteins (2/e)** Hames,
B.D. & Rickwood, D. (Eds)
...... Spiralbound hardback 0-19-963074-7 **$75.00**
...... Paperback 0-19-963075-5 **$50.00**
66. **Clinical Immunology** Gooi, H.C. & Chapel, H.
(Eds)
...... Spiralbound hardback 0-19-963086-0 **$69.95**
...... Paperback 0-19-963087-9 **$50.00**
65. **Receptor Biochemistry** Hulme, E.C. (Ed)
...... Spiralbound hardback 0-19-963092-5 **$70.00**
...... Paperback 0-19-963093-3 **$50.00**
64. **Gel Electrophoresis of Nucleic Acids (2/e)**
Rickwood, D. & Hames, B.D. (Eds)
...... Spiralbound hardback 0-19-963082-8 **$75.00**
...... Paperback 0-19-963083-6 **$50.00**
63. **Animal Virus Pathogenesis** Oldstone, M.B.A.
(Ed)
...... Spiralbound hardback 0-19-963100-X **$68.00**
...... Paperback 0-19-963101-8 **$40.00**
62. **Flow Cytometry** Ormerod, M.G. (Ed)
...... Paperback 0-19-963053-4 **$50.00**
61. **Radioisotopes in Biology** Slater, R.J. (Ed)
...... Spiralbound hardback 0-19-963080-1 **$75.00**
...... Paperback 0-19-963081-X **$45.00**
60. **Biosensors** Cass, A.E.G. (Ed)
...... Spiralbound hardback 0-19-963046-1 **$65.00**
...... Paperback 0-19-963047-X **$43.00**

59. **Ribosomes and Protein Synthesis** Spedding, G. (Ed)
...... Spiralbound hardback 0-19-963104-2 **$75.00**
...... Paperback 0-19-963105-0 **$45.00**
58. **Liposomes** New, R.R.C. (Ed)
...... Spiralbound hardback 0-19-963076-3 **$70.00**
...... Paperback 0-19-963077-1 **$45.00**
57. **Fermentation** McNeil, B. & Harvey, L.M. (Eds)
...... Spiralbound hardback 0-19-963044-5 **$65.00**
...... Paperback 0-19-963045-3 **$39.00**
56. **Protein Purification Applications** Harris,
E.L.V. & Angal, S. (Eds)
...... Spiralbound hardback 0-19-963022-4 **$54.00**
...... Paperback 0-19-963023-2 **$36.00**
55. **Nucleic Acids Sequencing** Howe, C.J. & Ward,
E.S. (Eds)
...... Spiralbound hardback 0-19-963056-9 **$59.00**
...... Paperback 0-19-963057-7 **$38.00**
54. **Protein Purification Methods** Harris, E.L.V. &
Angal, S. (Eds)
...... Spiralbound hardback 0-19-963002-X **$60.00**
...... Paperback 0-19-963003-8 **$40.00**
53. **Solid Phase Peptide Synthesis** Atherton, E. &
Sheppard, R.C.
...... Spiralbound hardback 0-19-963066-6 **$58.00**
...... Paperback 0-19-963067-4 **$39.95**
52. **Medical Bacteriology** Hawkey, P.M. & Lewis,
D.A. (Eds)
...... Spiralbound hardback 0-19-963008-9 **$69.95**
...... Paperback 0-19-963009-7 **$50.00**
51. **Proteolytic Enzymes** Beynon, R.J. & Bond, J.S.
(Eds)
...... Spiralbound hardback 0-19-963058-5 **$60.00**
...... Paperback 0-19-963059-3 **$39.00**
50. **Medical Mycology** Evans, E.G.V. &
Richardson, M.D. (Eds)
...... Spiralbound hardback 0-19-963010-0 **$69.95**
...... Paperback 0-19-963011-9 **$50.00**
49. **Computers in Microbiology** Bryant, T.N. &
Wimpenny, J.W.T. (Eds)
...... Paperback 0-19-963015-1 **$40.00**
48. **Protein Sequencing** Findlay, J.B.C. & Geisow,
M.J. (Eds)
...... Spiralbound hardback 0-19-963012-7 **$56.00**
...... Paperback 0-19-963013-5 **$38.00**
47. **Cell Growth and Division** Baserga, R. (Ed)
...... Spiralbound hardback 0-19-963026-7 **$62.00**
...... Paperback 0-19-963027-5 **$38.00**
46. **Protein Function** Creighton, T.E. (Ed)
...... Spiralbound hardback 0-19-963006-2 **$65.00**
...... Paperback 0-19-963007-0 **$45.00**
45. **Protein Structure** Creighton, T.E. (Ed)
...... Spiralbound hardback 0-19-963000-3 **$65.00**
...... Paperback 0-19-963001-1 **$45.00**
44. **Antibodies: Volume II** Catty, D. (Ed)
...... Spiralbound hardback 0-19-963018-6 **$58.00**
...... Paperback 0-19-963019-4 **$39.00**
43. **HPLC of Macromolecules** Oliver, R.W.A. (Ed)
...... Spiralbound hardback 0-19-963020-8 **$54.00**
...... Paperback 0-19-963021-6 **$45.00**
42. **Light Microscopy in Biology** Lacey, A.J. (Ed)
...... Spiralbound hardback 0-19-963036-4 **$62.00**
...... Paperback 0-19-963037-2 **$38.00**
41. **Plant Molecular Biology** Shaw, C.H. (Ed)
...... Paperback 1-85221-056-7 **$38.00**
40. **Microcomputers in Physiology** Fraser, P.J. (Ed)
...... Spiralbound hardback 1-85221-129-6 **$54.00**
...... Paperback 1-85221-130-X **$36.00**
39. **Genome Analysis** Davies, K.E. (Ed)
...... Spiralbound hardback 1-85221-109-1 **$54.00**
...... Paperback 1-85221-110-5 **$36.00**
38. **Antibodies: Volume I** Catty, D. (Ed)
...... Paperback 0-947946-85-3 **$38.00**
37. **Yeast** Campbell, I. & Duffus, J.H. (Eds)
...... Paperback 0-947946-79-9 **$36.00**
36. **Mammalian Development** Monk, M. (Ed)
...... Hardback 1-85221-030-3 **$60.00**
...... Paperback 1-85221-029-X **$45.00**
35. **Lymphocytes** Klaus, G.G.B. (Ed)
...... Hardback 1-85221-018-4 **$54.00**
34. **Lymphokines and Interferons** Clemens, M.J.,
Morris, A.G. & others (Eds)
...... Paperback 1-85221-035-4 **$44.00**
33. **Mitochondria** Darley-Usmar, V.M., Rickwood,
D. & others (Eds)
...... Hardback 1-85221-034-6 **$65.00**
...... Paperback 1-85221-033-8 **$45.00**

32. **Prostaglandins and Related Substances**
Benedetto, C., McDonald-Gibson, R.G. & others
(Eds)
.... Hardback 1-85221-032-X **$58.00**
.... Paperback 1-85221-031-1 **$38.00**
31. **DNA Cloning: Volume III** Glover, D.M. (Ed)
.... Hardback 1-85221-049-4 **$56.00**
.... Paperback 1-85221-048-6 **$36.00**
30. **Steroid Hormones** Green, B. & Leake, R.E.
(Eds)
.... Paperback 0-947946-53-5 **$40.00**
29. **Neurochemistry** Turner, A.J. & Bachelard, H.S.
(Eds)
.... Hardback 1-85221-028-1 **$56.00**
.... Paperback 1-85221-027-3 **$36.00**
28. **Biological Membranes** Findlay, J.B.C. & Evans,
W.H. (Eds)
.... Hardback 0-947946-84-5 **$54.00**
.... Paperback 0-947946-83-7 **$36.00**
27. **Nucleic Acid and Protein Sequence Analysis**
Bishop, M.J. & Rawlings, C.J. (Eds)
.... Hardback 1-85221-007-9 **$66.00**
.... Paperback 1-85221-006-0 **$44.00**
26. **Electron Microscopy in Molecular Biology**
Sommerville, J. & Scheer, U. (Eds)
.... Hardback 0-947946-64-0 **$54.00**
.... Paperback 0-947946-54-3 **$40.00**
25. **Teratocarcinomas and Embryonic Stem Cells**
Robertson, E.J. (Ed)
.... Hardback 1-85221-005-2 **$62.00**
.... Paperback 1-85221-004-4 **$0.00**
24. **Spectrophotometry and Spectrofluorimetry**
Harris, D.A. & Bashford, C.L. (Eds)
.... Hardback 0-947946-69-1 **$56.00**
.... Paperback 0-947946-46-2 **$39.95**
23. **Plasmids** Hardy, K.G. (Ed)
.... Paperback 0-947946-81-0 **$36.00**
22. **Biochemical Toxicology** Snell, K. & Mullock,
B. (Eds)
.... Paperback 0-947946-52-7 **$40.00**
19. **Drosophila** Roberts, D.B. (Ed)
.... Hardback 0-947946-66-7 **$67.50**
.... Paperback 0-947946-45-4 **$46.00**
17. **Photosynthesis: Energy Transduction** Hipkins,
M.F. & Baker, N.R. (Eds)
.... Hardback 0-947946-63-2 **$54.00**
.... Paperback 0-947946-51-9 **$36.00**
16. **Human Genetic Diseases** Davies, K.E. (Ed)
.... Hardback 0-947946-76-4 **$60.00**
.... Paperback 0-947946-75-6 **$34.00**
14. **Nucleic Acid Hybridisation** Hames, B.D. &
Higgins, S.J. (Eds)
.... Hardback 0-947946-61-6 **$60.00**
.... Paperback 0-947946-23-3 **$36.00**
13. **Immobilised Cells and Enzymes** Woodward, J.
(Ed)
.... Hardback 0-947946-60-8 **$0.00**
12. **Plant Cell Culture** Dixon, R.A. (Ed)
.... Paperback 0-947946-22-5 **$36.00**
11a. **DNA Cloning: Volume I** Glover, D.M. (Ed)
.... Paperback 0-947946-18-7 **$36.00**
11b. **DNA Cloning: Volume II** Glover, D.M. (Ed)
.... Paperback 0-947946-19-5 **$36.00**
10. **Virology** Mahy, B.W.J. (Ed)
.... Paperback 0-904147-78-9 **$40.00**

9. **Affinity Chromatography** Dean, P.D.G.,
Johnson, W.S. & others (Eds)
.... Paperback 0-904147-71-1 **$36.00**
7. **Microcomputers in Biology** Ireland, C.R. &
Long, S.P. (Eds)
.... Paperback 0-904147-57-6 **$36.00**
6. **Oligonucleotide Synthesis** Gait, M.J. (Ed)
.... Paperback 0-904147-74-6 **$38.00**
5. **Transcription and Translation** Hames, B.D. &
Higgins, S.J. (Eds)
.... Paperback 0-904147-52-5 **$38.00**
3. **Iodinated Density Gradient Media** Rickwood,
D. (Ed)
.... Paperback 0-904147-51-7 **$36.00**

Sets

**Essential Molecular Biology: Volumes I and II
as a set** Brown, T.A. (Ed)
.... Spiralbound hardback 0-19-963114-X **$118.00**
.... Paperback 0-19-963115-8 **$78.00**
Antibodies: Volumes I and II as a set Catty, D.
(Ed)
.... Paperback 0-19-963063-1 **$70.00**
Cellular and Molecular Neurobiology Chad, J.
& Wheal, H. (Eds)
.... Spiralbound hardback 0-19-963255-3 **$133.00**
.... Paperback 0-19-963254-5 **$79.00**
**Protein Structure and Protein Function: Two-
volume set** Creighton, T.E. (Ed)
.... Spiralbound hardback 0-19-963064-X **$114.00**
.... Paperback 0-19-963065-8 **$80.00**
DNA Cloning: Volumes I, II, III as a set
Glover, D.M. (Ed)
.... Paperback 1-85221-069-9 **$92.00**
**Molecular Plant Pathology: Volumes I and II as
a set** Gurr, S.J., McPherson, M.J. & others (Eds)
.... Spiralbound hardback 0-19-963354-1 **$0.00**
.... Paperback 0-19-963353-3 **$0.00**
**Protein Purification Methods, and Protein
Purification Applications, two-volume set**
Harris, E.L.V. & Angal, S. (Eds)
.... Spiralbound hardback 0-19-963048-8 **$98.00**
.... Paperback 0-19-963049-6 **$68.00**
**Diagnostic Molecular Pathology: Volumes I and
II as a set** Herrington, C.S. & McGee, J. O'D.
(Eds)
.... Spiralbound hardback 0-19-963241-3 **$0.00**
.... Paperback 0-19-963240-5 **$0.00**
**Receptor Biochemistry; Receptor-Effector
Coupling; Receptor-Ligand Interactions**
Hulme, E.C. (Ed)
.... Spiralbound hardback 0-19-963096-8 **$193.00**
.... Paperback 0 19-963097-6 **$125.00**
Signal Transduction Milligan, G. (Ed)
.... Spiralbound hardback 0-19-963296-0 **$60.00**
.... Paperback 0-19-963295-2 **$38.00**
**Human Cytogenetics: Volumes I and II as a set
(2/e)** Rooney, D.E. & Czepulkowski, B.H. (Eds)
.... Hardback 0-19-963314-2 **$130.00**
.... Paperback 0-19-963313-4 **$90.00**
**Peptide Hormone Secretion/Peptide Hormone
Action** Siddle, K. & Hutton, J.C. (Eds)
.... Spiralbound hardback 0-19-963072-0 **$135.00**
.... Paperback 0-19-963073-9 **$90.00**

ORDER FORM for USA and Canada

Qty	ISBN	Author	Title	Amount
			S&H	
	CA and NC residents add appropriate sales tax			
			TOTAL	

Please add shipping and handling: $2.50 for first book, ($1.00 each book thereafter)

Name ....

Address ....

....

.... Zip ....

[ ] Please charge $ .... to my credit card
Mastercard/VISA/American Express (circle appropriate card)

Acct. .... Expiry date ....

Signature ....

Credit card account address if different from above:

....

.... Zip ....

[ ] I enclose a cheque for $...........

Mail orders to: Order Dept. Oxford University Press, 2001 Evans Road, Cary, NC 27513